Štefan Palágyi
Karel Štamberg
Dušan Vopálka

Determination of transport parameters of radionuclides

AF141883

Štefan Palágyi
Karel Štamberg
Dušan Vopálka

Determination of transport parameters of radionuclides

LAP LAMBERT Academic Publishing

Impressum / Imprint
Bibliografische Information der Deutschen Nationalbibliothek: Die Deutsche Nationalbibliothek verzeichnet diese Publikation in der Deutschen Nationalbibliografie; detaillierte bibliografische Daten sind im Internet über http://dnb.d-nb.de abrufbar.
Alle in diesem Buch genannten Marken und Produktnamen unterliegen warenzeichen-, marken- oder patentrechtlichem Schutz bzw. sind Warenzeichen oder eingetragene Warenzeichen der jeweiligen Inhaber. Die Wiedergabe von Marken, Produktnamen, Gebrauchsnamen, Handelsnamen, Warenbezeichnungen u.s.w. in diesem Werk berechtigt auch ohne besondere Kennzeichnung nicht zu der Annahme, dass solche Namen im Sinne der Warenzeichen- und Markenschutzgesetzgebung als frei zu betrachten wären und daher von jedermann benutzt werden dürften.

Bibliographic information published by the Deutsche Nationalbibliothek: The Deutsche Nationalbibliothek lists this publication in the Deutsche Nationalbibliografie; detailed bibliographic data are available in the Internet at http://dnb.d-nb.de.
Any brand names and product names mentioned in this book are subject to trademark, brand or patent protection and are trademarks or registered trademarks of their respective holders. The use of brand names, product names, common names, trade names, product descriptions etc. even without a particular marking in this work is in no way to be construed to mean that such names may be regarded as unrestricted in respect of trademark and brand protection legislation and could thus be used by anyone.

Coverbild / Cover image: www.ingimage.com

Verlag / Publisher:
LAP LAMBERT Academic Publishing
ist ein Imprint der / is a trademark of
OmniScriptum GmbH & Co. KG
Heinrich-Böcking-Str. 6-8, 66121 Saarbrücken, Deutschland / Germany
Email: info@lap-publishing.com

Herstellung: siehe letzte Seite /
Printed at: see last page
ISBN: 978-3-659-76841-5

Determination of transport parameters of radionuclides in granular materials using simple model approach

Š. Palágyi, K. Štamberg and D. Vopálka

Department of Nuclear Chemistry, Faculty of Nuclear Sciences and Physical Engineering, Czech Technical University in Prague, 115 19 Prague, Czech Republic

E-mail: istvan.palagyi@email.cz

Keywords

Columns · Sorption and desorption · Groundwater · Pulse and step inlets · Dynamic modeling · 1-D ADE equation · Breakthrough curves

Table of Contents

Abstract

The assessment of the ability of natural barriers to retain radionuclides and retard their transfer in groundwater requires knowledge of important transport parameters, the retardation and dispersion coefficients. The use of dynamic techniques is in this task more effective than that of batch technique, as the conditions of dynamic experiments better simulate the real systems, in which the contaminated groundwater is flowing through the bed of a porous (grained) solid material (crushed rock, soil, or sediment). Two techniques of the contaminant inlet, the pulse injection and step (continuous) inlet are obviously applied. Dynamic column experiments make possible to study the influence of sorption or desorption of studied contaminants on the velocity of their transport through the saturated or unsaturated bed. The transport parameters are determined in the course of evaluation of experimental data, which generally consists of the regression of breakthrough curve by selected analytical solution of the 1-D advection-dispersion equation (ADE). With the respect to the kinetics of the contaminant interaction with the surface of the solid phase, there are two basic groups of these solutions: the first responds to the equilibrium dynamics, and the second one to so-called non-equilibrium dynamics. In description of interaction, that implies the mathematical form of the solution of transport equation, it is further possible to specify both the equilibrium isotherm (linear or non-linear) and the type of kinetic equation (e.g., linear driving force (LDF) model). In this paper, a set of simplified equilibrium dynamic models is presented, that could be recommended for the evaluation of an important range of column experiment in heterogeneous systems accomplished under the equilibrium dynamics conditions.

Preface

This booklet is a summary of our work in the past ten years in the study of transport phenomena in granular materials and groundwater systems carried out in the Nuclear Research Institute at Rez and Department of Nuclear Chemistry of Czech Technical University in Prague. The main goal of our work was to find simple methods of mathematical modeling, which make possible to determine each of significant transport parameters of radionuclides in order to predict their behavior in studied systems. The presented examples of the use of this method are only partially taken from our previous work in this area. The full papers can be obtained in References. The data obtained by this simple method may well serve in the depiction of far field environment of the radioactive waste depository under normal conditions. In spite of fact that for some of us this work would be the swan-song of a relatively short period, we are convinced in the necessity of further studying this method mainly for the system, which have not been anticipated in the preceding studies.

1. Introduction

The assessment of the final disposal of high-level radioactive waste in geological formations requires studies on transport of critical radionuclides (e.g., ^{135,137}Cs, ^{89,90}Sr, ^{129,131}I and ^{75,79}Se) in the surrounding environment, with the aim to predict, in case of the failure of engineering barriers of a repository, the possible effect of released radionuclides presence in the groundwater, as a source of drinking or irrigation water [1,2]. For the evaluation of the ability of the environment of nuclear installations, namely their natural barriers, to retain radionuclides and prevent them to transfer into groundwater, knowledge of certain transport parameters is necessary. The most important amongst them are the retardation and dispersion (or *Pe*-number) coefficients. If equilibrium dynamics and linear equilibrium isotherms can be supposed, the retardation coefficient is constant and can be calculated using the value of the distribution coefficient (K_d) determined by batch test, which is a convenient laboratory method. However, if non-linear equilibrium isotherm holds [2-6], the distribution coefficient is not constant and the equilibrium isotherm has to be

2

described as a function of actual equilibrium concentration in the liquid phase, e.g., by two-parametric Freundlich equation, and as a result, the retardation coefficient is then also a function of the concentration in the liquid phase moving through the bed (column).

For the determination of transport parameters, it is more effective to use the dynamic column (transport) process, where the groundwater flows through the bed (in column) of a porous (grained) solid material (e.g., crushed crystalline rock, soil, or sediment) and the so-called breakthrough curve (BTC) is obtained. According to the input technique, there are two possible forms of BTC: (i) bell-curve, if pulse tracer input is used, (ii) S-curve, if step (continuous) tracer input of liquid phase is realized. The determination of transport parameters, mentioned above, is performed by the fitting of experimental breakthrough curve using the suited model. Generally, BTC is the function of a relative radioactivity, or molar concentration of given component on the number of pore volumes at the column outlet, depending on the type of sorption and desorption equilibrium isotherms, sorption/desorption kinetics, type of dynamics, i.e., equilibrium or non-equilibrium dynamics, and physical properties of the bed in column (e.g., grain size distribution, shape of grains, character of the flow, channeling, dispersion), as well [7, 8].

Dynamic column experiments make possible to observe the sorption and/or desorption during transport of a contaminant by water through the bed under saturated or unsaturated conditions. There are a lot of advantages of the column operation (open arrangement) in comparison with the batch one (closed arrangement). Above all, the dynamic arrangement is much closer to conditions existing in the environment because the aqueous volume to mass ratio is more realistic, the build-up of interfering products is greatly limited, the change of material quality during test is insignificant, there is a possibility to apply undisturbed materials (core samples)[9] that enables to keep a fine structure and specificity of the sample, and allows easier differentiation of reversible and

3

irreversible portions of sorbed and desorbed contaminants [3-6]. On the contrary, the advantages of static (batch) method lie in a much more simple experimental apparatus and the possibility to attain for certain radionuclide an equilibrium distribution state. The column technique is effective mainly in such types of systems, in which the advection transport of the given radionuclide is dominant over their diffusion, or interaction-dependent movement [3,4,6,10,11].

If for the given system the linear equilibrium isotherm holds, it has to be added that the retardation coefficient (R_{exp}) can be evaluated directly from the maximum on the bell-shaped curve of BTC, or from the point of inflexion on S-curve, and the determination of distribution coefficient (K_d) can be calculated immediately. As for the hydrodynamic dispersion coefficient (D_d), it can be determined by separate laboratory column technique, as well.

Recently some very interesting papers concerned with the modeling of sorption/desorption processes have been published [12-14]. At first, the review summarizing the modeling of fixed-bed adsorption in aqueous systems [12] has to be taken into consideration. Not only the single-component equilibrium isotherm models (e.g., Freundlich-, Langmuir-isotherm, and six others) are dealt with, but also multi-component ones (five models) are discussed. This paper includes also the non-equilibrium dynamic models: (i) so-called general rate models with six types of different interaction-kinetic terms (with, e.g., macropore and micropore diffusion in series/in parallel, shrinking core model, non-uniform size of particles, concentration-dependent surface diffusion coefficient), (ii) linear-driving force (LDF) model, (iii) models of Clark, Thomas, Bohardt-Adams, Yoon-Nelson, Wang and others, (iv) then models based on wave propagation and constant pattern theory. Also, the very important basic correlations can be found here. They deal with relations based on non-dimensional numbers type of Sc, Re, Pe and Sh by means of which the parameters (as bulk liquid diffusivity, surface film (external) diffusivity, hydrodynamic dispersion coefficient, different types of diffusion coefficients) can be calculated.

Six types of different equilibrium isotherm models (again, Langmuir, Freundlich, and so on) and two types of non-equilibrium dynamic models (the classical, so-called mass transfer model involving the Langmuir isotherm and LDF kinetic function, and the Clark model) are a subject of paper [13]. In the course of calculation of breakthrough curves, it was found that the application of batch isotherm parameters can result in inaccurate estimations and therefore, the so called continuous isotherm parameters, obtained from the experimental column data, should be used. The finite difference method, described there in detail, was used for the numerical solution of differential equation systems.

The simplified approach was applied in the study of vertical concentration profile in cores of marine sediment [14] based on the application of the compartment model, and of the Green function obtained by analytical solution of the diffusion-advection equation for conditions of pulse tracer inlet (injection). The advantage of the compartment model consists in its simplicity, no detailed input data are necessary, it deals with the black-box model. The Green function is applicable without problems only in the special cases. This type of the solution is fast, but – on the other hand – it does not include the interaction parameters and therefore, it should not be used for the description (modeling) of the sorption or desorption systems, as it is evident from the figures presented in paper [14]. The Green functions, again without of interaction parameters can also be found in paper [15], where the theoretical description of diffusion and migration of ^{137}Cs in soil is presented.

The new approach to the construction of the non-equilibrium dynamic model, relating to the sorption in packed-bed columns and supposing the non-linear equilibrium isotherm of the Langmuir type, and LDF kinetic model, is described in [16]. Really, it deals with the sophisticated derivation of a new mathematical model that significantly improves the prediction of the shape of breakthrough curves. It is based on a dual porosity model combining external mass transfer and intra particle transport by solid (surface) diffusion, and assumes a parabolic concentration profile within the particle. The finite difference method by means of implicit scheme was used to the solution of given differential equations.

In this paper we shall give a review of the most widely used column techniques and evaluations of results obtained by them. The evaluations of column experiments are obviously based on the application of relations resulting from the analytical solution of a 1-D advection-dispersion equation (ADE, including the interaction term), namely for both pulse and step (continuous) tracer input of a radioactive solute. From the mathematical point of view, these relations – dependent on the initial and boundary conditions or on the type of solute input – are in the form of exponential or complementary error (erfc) functions. In principle, this review is focused on such type of models (relations), the advantage of which consists in their simplicity and fast mathematical solubility, which are important properties in a case of modeling and prediction column sorption/desorption processes, especially if it deals with evaluation of laboratory experiments.

It is shown that the methods (models) of evaluation can be used in the case of validity of linear or a non-linear sorption/desorption equilibrium isotherm, and that the experimental data can be fitted by means of a Newton-Raphson multidimensional non-linear regression procedure, in which the regression function is based on the above mentioned relations. Values of basic parameters are sought in the course of the regression procedure [17-20]. Applicability of the method has been demonstrated on the evaluation of BTCs of the several environmentally important radionuclides obtained by column experiments of various crushed granite and disturbed soil beds and undisturbed soil cores, where radionuclide spiked synthetic groundwater was used [6,8,21-25]. A concise review of our work in this area of study has been recently reported in Ref. [26].

2. Simplified sorption/desorption equations and their application

2.1. Pulse tracer inlet of radioactive contaminants

2.1.1. Radioactive contaminants transport modeling

The equilibrium dynamics of the sorption/desorption process in the 1-D steady state flow can be described by the 2nd order partial differential equation (1) including the interaction of given radionuclide with solid phase and its radioactive decay, characterized by retardation coefficient (R) and decay constant (λ), respectively. It deals with the classical advection-dispersion equation (ADE) [20,27]:

$$\frac{\partial C}{\partial t} = \frac{D_d}{R_{cal}} \frac{\partial^2 C}{\partial x^2} - \frac{u}{R_{cal}} \frac{\partial C}{\partial x} - \frac{\lambda}{R_{cal}} \left(C + \frac{\rho}{\theta} f(C) \right) + \dots \tag{1}$$

Solution of this equation, following necessary substitutions, simplifications, e.g., if radioactive decay is negligible, and initial [$C_t = 0$, and $x = 0$ at $t_0 = 0$] and boundary [$C\,(0,t)$, $x(0,L)$] conditions yields the analytical form (Eq. (2)), which can well be used for fitting a real, bell-shaped, by pulse tracer inlet obtained breakthrough curve (BTC). Namely, as a dependence of $C_{rel} = f(n_{PV})$ or $A_{rel} = f(n_{PV})$, where C_{rel} or A_{rel} represent the relative concentration or the relative radioactivity of a given component, [8,20,21]:

$$C_{rel} = \frac{C}{C_0} = \frac{k_{int} R_{cal}}{L. \sqrt{\dfrac{4\pi\, R_{cal}\, n_{PV}}{Pe}}} e^{-\dfrac{Pe}{4\,R_{cal}\,n_{PV}}(R_{cal} - n_{PV})^2} \tag{2}$$

A_{rel} ($= A/A_0$) can be used instead of C_{rel}.

If the linear isotherm holds in the system, K_d is constant, then R_{cal} is constant, too (see Eq. (3)):

$$R_{cal} = 1 + \frac{\rho}{\theta} K_d \quad , \tag{3}$$

where K_d ($= q/C$) is the distribution coefficient (q is the equilibrium concentration of investigated component in the solid phase (mol/kg) and C in the liquid phase (mol m^{-3}).

If the isotherm is non-linear, $K_d \neq$ const., than R_{cal} is a function of the concentration, namely, of the first derivation of the equilibrium isotherm in point C [20]:

$$R_{cal} = 1 + \frac{\rho}{\theta} \frac{dq}{dC} \tag{4}$$

For Freundlich type of non-linear sorption isotherm can be written [20]:

$$q = k_F \cdot C^{n_F} \tag{5}$$

$$\frac{dq}{dC} = n_F \cdot k_F \cdot (C_{rel} \cdot C_0)^{n_F - 1} \tag{6}$$

$$R_{cal} = 1 + (\rho / \theta) \cdot n_F \cdot k_F \cdot (C_{rel} \cdot C_0)^{n_F - 1} \tag{7}$$

In Eqs. (1)-(7) it means: C or A – concentration or radioactivity of the investigated component in the liquid phase in the outlet from the column, C_0 or A_0 - initial liquid phase concentration or radioactivity in the inlet to the column, R_{cal} - retardation coefficient (dimensionless), D_d – hydrodynamic dispersion coefficient, i.e., coefficient of axial dispersion (m^2 s^{-1}), u – linear seepage velocity (flow-rate) through the bed void cross-section (m s^{-1}), ρ – bulk density of the solid phase (kg m^{-3}), θ – bed porosity (m^3 m^{-3}), x - longitudinal (axial) bed coordinate (m), λ – radioactive decay constant (s^{-1}), L – column length, i.e., bed height (m), k_{int} – integration constant (m) equals approx. to column length, $n_{PV} = \sum V/PV$ – the number of pore volumes of liquid phase entered in the column, where PV is one pore volume and $\sum V$ is the total volume of the liquid phase at the outlet from the column, $Pe = (u \cdot L / D_d)$ dimensionless Peclet number, k_F - the sorption capacity coefficient (m^3 kg^{-1}) and n_F – sorption intensity coefficient (dimensionless), both characterizing the Freundlich equation.

As it was mentioned above, the geometrical form of Eq. (2) is a bell-shaped curve. At the maximum, i.e., at the peak of the breakthrough curve ($C_{rel,max}$), the corresponding number of pore volumes equals numerically to the retardation coefficient, therefore: $R_{exp} = (n_{PV})_{max}$. (But, it is valid for equilibrium linear isotherm only, when $q = K_d \cdot C$.)

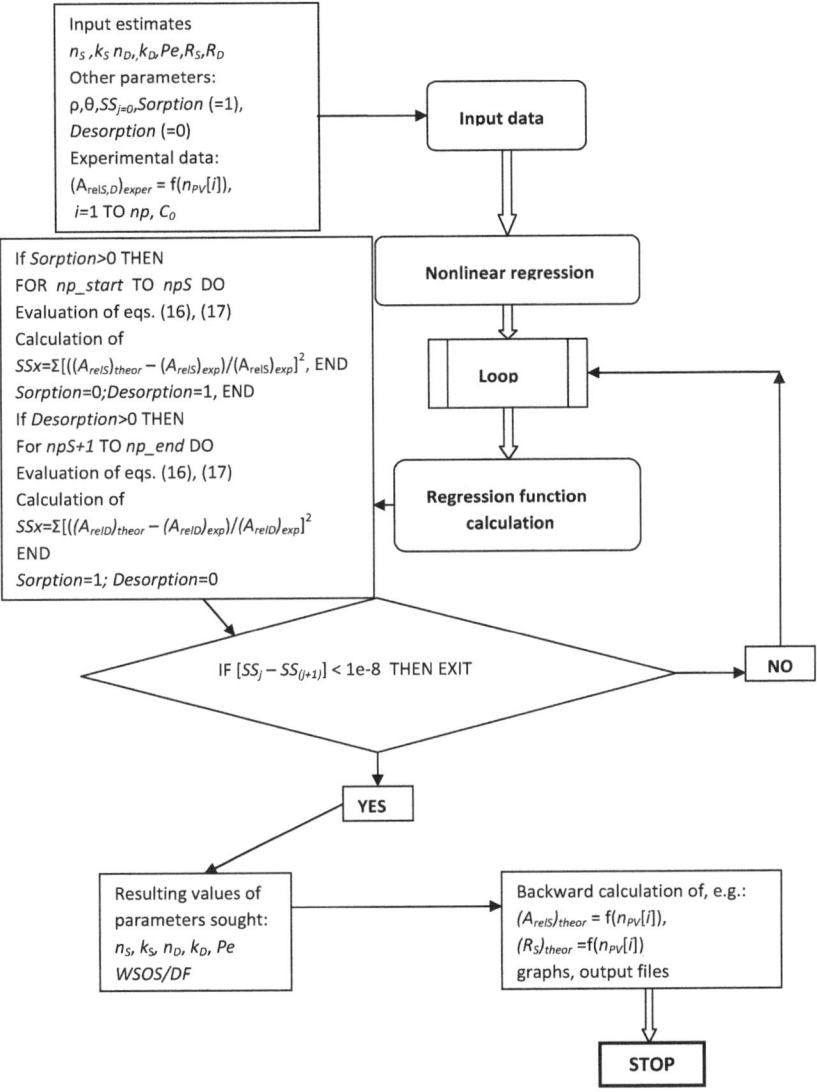

Fig. 1 Algorithm (flow sheet) of the PNLRPal1.fm program. From Palágyi Š and Štamberg K (2011) Cent Eur J Chem 9:798. With permission

The procedure for evaluation of experimental data, can be based on the Newton-Raphson multidimensional method of non-linear regression [19], by means of which the data are fitted using a regression function, e.g., consisting of the Eqs. (2) and (4), mentioned above. Simultaneously, the following parameters of the ADE equation are searched:

(1) The equilibrium isotherm parameters (e.g., of the Freundlich isotherm, n_F and k_F) by means of which, and known hydraulic parameters (ρ and θ), the retardation coefficient, R_{cal}, is calculated (see Eqs. (4)-(7)).

(2) The Pe number on the base of which, and known seepage velocity (u) and column length (L), hydrodynamic dispersion coefficient, D_d ($= u \cdot L/Pe$), can be determined.

(3) The integration constant k_{int} - approximately equal to L, as it was also confirmed experimentally [20].

The fitting proceeds in the iteration cycle (which is described in detail in several our papers [22,23]), from which it is possible to withdraw the sought parameters when the difference of the sum of relative squares of deviations $(SSx)_i$ after two successive cycles (i.e., i_{th} and i_{th+1}) is less than 10^{-8}:

$$ (SSx)_i = \sum_{i=1}^{N} \left(\frac{\left(A_{rel,cal} - A_{rel,exp} \right)_i}{\left(A_{rel,exp} \right)_i} \right)^2 \tag{8} $$

The respective computational code PNLRPal5.fm (Code Package Stamb-2008) is constructed (Fig. 1) for the FAMULUS software product [17]. As fitting criterion, reflecting the agreement between calculated and experimental values, the χ^2 defined by Eq. (9) is used, and subsequently, the criterion of goodness-of-fit, $WSOS/DF$ (Weighted Sum Of Squares divided by the Degrees of Freedom), is calculated by Eq. (10) [18]:

$$ \chi^2 = \sum_{i=1}^{N} \frac{(SSx)_i}{(s_q)_i^2} \tag{9} $$

$$ \frac{WSOS}{DF} = \frac{\chi^2}{n_i} , \quad n_i = n_p - n \tag{10} $$

where: s_q – standard uncertainty of experimental determination of $A_{rel,exp}$, N, n_p – overall number of experimental points, n_i – number of degrees of freedom, n – number of searched parameters (in our case $n = 4$). The agreement is acceptable if $0.1 \leq WSOS/DF \leq 20$ [18].

2.1.2. Evaluation of pulse tracer inlet experimental data

If it is supposed, that the equality relation $k_{int} = L$ holds, Eq. (11) is obtained from Eq. (2):

$$A_{rel} = \frac{A_n}{A_{rel,max}} = \frac{R_{cal}}{\sqrt{\dfrac{4\pi\,R_{cal}\,n_{PV}}{Pe}}}\, e^{-\frac{Pe}{4\,R_{cal}n_{PV}}(R_{cal}-n_{PV})^2} \quad, \tag{11}$$

and this equation was used in fitting experimental BTCs by theoretical BTCs in subchapters 2.1.2.1 and 2.1.2.2. In further experiments it was shown that the requirement of $R_{theor} = R_{exp}$ to be fulfilled needs further modifications of Exp. (11). Therefore, the Eq. (11) was modified [20,21], namely by means of the so called: (i) peak position correction coefficient, k_p, (ii) and height correction coefficient, k_h. The following corresponding relations (see Eqs (12) and (13)) are derived and inserted into Eq. (11). (As mentioned above, $R_{exp} = (n_{PV})_{max}$ at $A_{rel,max}$ or $C_{rel,max}$.).

$$k_p = \frac{R_{cal}}{R_{exp}} \quad, \tag{12}$$

$$k_h = \sqrt{\frac{k_p Pe}{4\pi}}\, e^{-\frac{Pe\,(k_p-1)^2}{4k_p}} \quad. \tag{13}$$

It is evident that if k_p converges to unity then $k_h \rightarrow 0.5\,(Pe/\pi)^{0.5}$. The result of above mentioned corrections is that general form of Eq. (11) should be rewritten as Eq. (14), where $A_{rel,cor}$ is the corrected A_{rel}:

$$A_{rel,cor} = \frac{k_p}{k_h}\, \frac{R_{exp}}{\sqrt{\dfrac{4\pi\,k_p R_{exp}\,n_{PV}}{Pe}}}\, e^{-\frac{Pe}{4\,k_p R_{exp}n_{PV}}(k_p R_{exp}-n_{PV})^2} \tag{14}$$

This equation can be used for fitting of BTC experimental points including linear or non-linear isotherm approach of sorption/desorption process as in subchapters 2.1.2.3 and 2.1.2.4.

2.1.2.1. Migration of ^{137}Cs and ^{152,154}Eu in crushed crystalline rocks

As the first example the results of study of ^{137}Cs and ^{152,154}Eu transport in crystalline rocks may serve [8]. The crushed samples of five rocks, diorite-I, diorite-II, gabbro, granite, and tonalite, were dry sieved and a fraction of 0.25–0.80 mm was used in the research. The samples originated from Cavernous Gas Reservoir near Příbram (Czech Republic). The selected crystalline granitic rocks are a part of the Central Bohemian Pluton. As columns, commercially available translucent 20 cm^3 PE injection syringes of (2.1 cm in I.D. and 8.8 cm in length) with tapered outlet were used. Pore-water volumes were 11.0-11.4 cm^3. The equipment (Fig. 2) consisted of 4 columns, each connected to 4 peristaltic single channel pumping heads (Easy Load II L/S Model No. 77201-60) on the inlet. The pumping heads were mounted on a common shaft to one Masterflex Console Drive, Model No. 7520-57 (220 VA), and connected to four 1-dm^3 glass reservoirs wrapped into tin foil to prevent developing of algae.

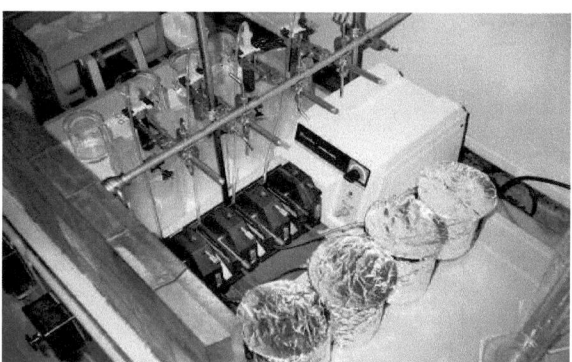

Fig. 2 Apparatus used in transport experiments (From Palágyi Š et al. (2009) J Radioanal Nucl Chem 279:431. With permission)

The fractions of the outflow were collected in 1 dm^3 covered beakers. The 0.3 cm^3/min of liquid flow rate was kept constant throughout the experiments. The water phase was introduced from the reservoir into the top of the column. In the sorption experiments, artificial (synthetic) groundwater (SGW) of Na-hydrogen-carbonate and Na-carbonate type was used, of pH approximately 8.5 and $0.83 \cdot 10^{-2}$ mol/dm^3 ionic strength. The amount of solution of radionuclides was about 0.1 cm^3 (10–20 kBq). Carrier-free $^{137}Cs^+$ and $^{152,154}Eu^{3+}$ were used as nitrates of high radiochemical and radionuclidic purity (min. 99%). Carrier CsNO$_3$ and Eu(NO$_3$)$_3$ in concentration of 10^{-6} mol/dm^3 were added to the synthetic groundwater. Desorption experiments followed the sorption experiment directly, mixture of a 2:1 (v/v) H$_2$SO$_4$ and HNO$_3$ was used, the first in 10^{-4} N and then in 10^{-2} N concentration, simulating acid rain.

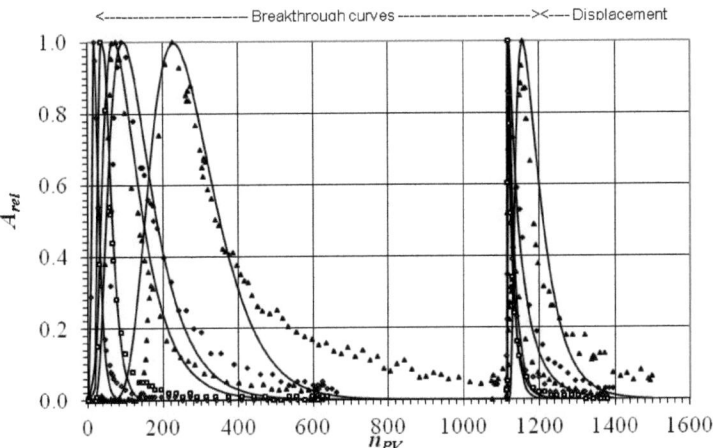

Fig. 3 Experimental (symbols) and theoretical (solid lines) BTCc of ^{137}Cs sorption and desorption on diorite-I (▲), diorite-II (♦), gabbro (Δ), granite (◊) and tonalite (□) (From Palágyi Š et al. (2009) J Radioanal Nucl Chem 279:431. With permission)

Sorption and desorption (displacement) breakthrough curves of ^{137}Cs in all five crushed rocks were obtained, as seen in Fig. 3. The BTCs can be characterized generally by relatively long tails, indicating that ^{137}Cs sorption to the rocks grains is, at least in part, controlled by rate-limited reactions [28]. Therefore, the theoretical BTCs, based on linear isotherm approach, in tailing part differ from experimental BTCs. Nevertheless, as it is shown further, this fact does not affect considerably the values of the transport parameters for mobile fractions (Table 1).

Table 1 The most important experimental and theoretical transport parameters of the sorption of radionuclides in crushed rocks (From Palágyi Š et al. (2009) J Radioanal Nucl Chem 279:431. With permission)

Crushed rocks	Retardation coefficient		Distribution coefficient		$K_{d,s \text{(theor)}}$: $K_{d,s \text{(exp)}}$	Pe number	Hydrodynamic dispersion coefficient
	$R_{s \text{(exp)}}$ (-)	$R_{s \text{(theor)}}$ (-)	$K_{d,s \text{(exp)}}$ (cm^3/g)	$K_{d,s \text{(theor)}}$ (cm^3/g)	(-)	(-)	$D_{d,s}$ (cm^2/min)
				^{137}Cs			
Diorite-I	227	245	91	98	1.1	15	0.09
Diorite-II	90	110	31	38	1.2	6	0.23
Gabbro	70	85	26	32	1.2	5	0.26
Granite	17	19	6	7	1.1	8	0.16
Tonalite	36	43	12	15	1.2	12	0.12
			Continuation				
			^{152,154}Eu[1,2]				
Diorite-I	555	555	223	223	1.0	ND	ND
Diorite-II	562	562	193	193	1.0	ND	ND
Gabbro	530	530	197	197	1.0	ND	ND
Granite	536	536	189	189	1.0	ND	ND
Tonalite	550	550	185	185	1.0	ND	ND

[1] Estimated R_s values for each rock because of the absence of respective BTCs.
[2] ND - Not determined.

The theoretical curves, calculated according to Eq. 11, and the measured points agree well mainly in the peak areas, from which the principal transport parameters are derived, namely the retardation coefficients. Further, the activity balance made for each rock has shown that only 7–9% of the eluted activities with SGW lie outside the theoretical breakthrough fitting curve in the case of diorite-I, diorite II, gabbro and tonalite. In the case of granite this eluted activity is below 1%. The difference between experimental and theoretical values on the descending part of BTC is greater than 1% in diorite-I (<4%) and diorite-II (<2%), only. Therefore, the obtained results are fully suitable mainly for mutual comparison of the sorption ability of the studied rocks. Eu exhibits a complete different sorption behavior. The prevailing part of the 152,154Eu activity is fixed (immobilized) in the column material, which under the given conditions can hardly be desorbed by SGW and is in practice permanently bound to the rock constituents. Only about 13% of total europium was mobile and could be desorbed.

2.1.2.2. Sorption and desorption of $^{125}I^-$, $^{137}Cs^+$, $^{85}Sr^{2+}$ and $^{152,154}Eu^{3+}$ on soils

Next example of radionuclide transport is the sorption and desorption of $^{125}I^-$, $^{137}Cs^+$, $^{85}Sr^{2+}$ and $^{152,154}Eu^{3+}$ on disturbed soils [6]. The samples were taken at several locations from $5-25$ cm (S-3/1) and from $80-100$ cm (S-3/4) depths. The dry samples were gently pulverized and get rid of small stones by sieving through a 2.5 mm sieve. The synthetic groundwater (SGW) was the same as that in previous subchapter.

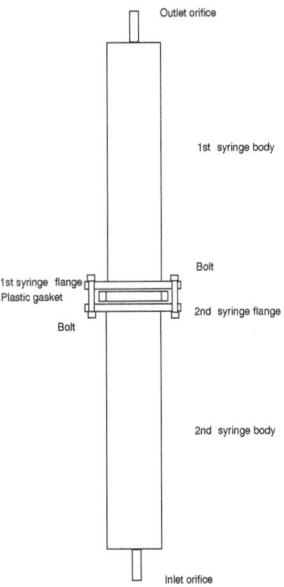

Fig. 4 Column arrangement used in experiments with disturbed soil samples (From Palágyi Š and Vodičková H (2009) J Radioanal Nucl Chem, 280:3. With permission

A pair of commercially available 20 cm^3 PP+PE injection syringes with tapered outlet and removed plungers placed against each other were used to construct the columns. The bodies of two syringes with flanges were joined through specially manufactured gasket and fixed with 6 bolts (Fig. 4). The resulted 2.1 cm inner diameter columns were 17.8 cm in length. Pore-water volumes were 26.2-33.6 cm^3. Carrier-free ^{125}Γ, ^{137}Cs$^+$, ^{85}Sr^{2+} and ^{152,154}Eu^{3+} of high radiochemical and radionuclidic purity (min. 99%) were used (several tens of kBq). The water-soluble compounds (NaI and Cs, Sr and Eu nitrates) were added into SGW as carriers in 10^{-6} mol/dm^3 concentrations. The SGW was introduced from the reservoirs into the columns at the bottom, i.e., against gravity. The SGW with constant flow-rate of ca 0.1 cm^3/min, (i.e., about 0.060±0.005 cm/min of seepage velocity) was introduced from the reservoirs into the columns at the bottom. Solutions with radionuclides were applied individually into the stream of SGW entering the columns in the amount of 0.1 cm^3 using a 1-cm^3 syringe equipped with a sharp needle by perforating the inlet tubing.

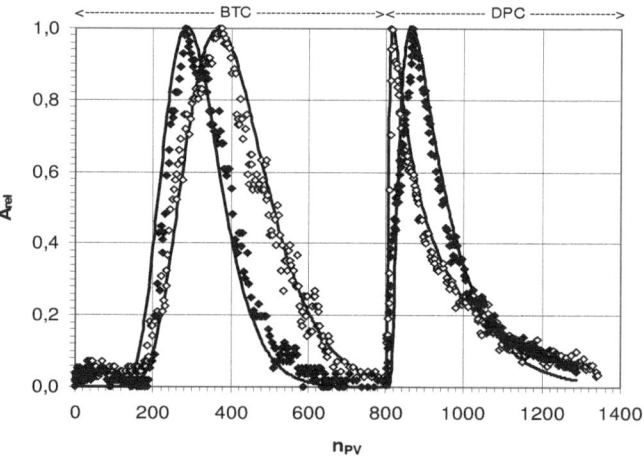

Fig. 5 Experimental (symbols) and theoretical (solid lines) breakthrough curves of ^{137}Cs dynamic sorption on S-3/1 (◊) and S-3/4 (♦) soil samples (From Palágyi Š and Vodičková H (2009) J Radioanal Nucl Chem, 280:3. With permission)

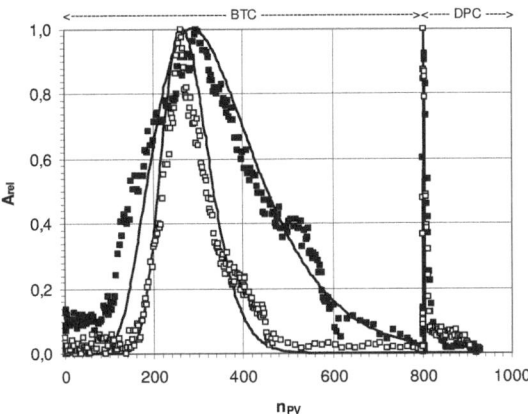

Fig. 6 Experimental (symbols) and theoretical (solid lines) breakthrough curves of ^{85}Sr dynamic sorption on S-3/1 (□) and S-3/4 (■) soil samples (From Palágyi Š and Vodičková H (2009) J Radioanal Nucl Chem, 280:3. With permission)

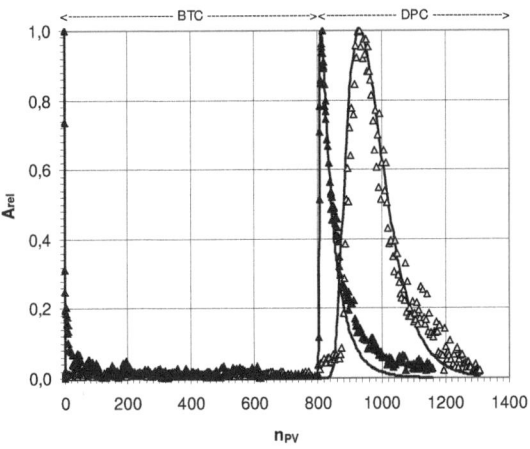

Fig. 7 Experimental (symbols) and theoretical (solid lines) breakthrough curves of $^{152,154}Eu$ dynamic sorption of S-3/1 (△) and S-3/4 (▲) soil samples (From Palágyi Š and Vodičková H (2009) J Radioanal Nucl Chem, 280:3. With permission)

BTC of sorption and displacement (DTC) experiments of ^{137}Cs, ^{85}Sr and $^{152,154}Eu$ are presented in Fig. 5-7 (^{125}I was omitted as no sorption occurs) and results obtained by fitting are presented in Table 2. It should be noted that displacement (desorption) experiments is related to the fraction remained on soil samples after sorption. It can be seen that the theoretical BTCs more or less follows the experimental BTCs, if some correction factors are used $k_c \approx 1$. Linear isotherm approach is used.

Table 2 The most important experimental and theoretical transport parameters of the dynamic sorption and desorption of radionuclides in disturbed surface and subsurface soils (From Palágyi Š and Vodičková H (2009) J Radioanal Nucl Chem, 280:3. With permission)

Radio nuclides	Soil samples	Sorption		Displacement		k_c	Pe	$D_{d,s}$
		$R_{s,exp}$	$R_{s,theor}$	$R_{ds,exp}$	$R_{ds,theor}$	$R_{s,exp}/R_{s,theor}$	-	cm^2/min
^{125}I	S-3/1	0.6	0.7	0.6	0.7	0.86	6	0.13
	S-3/4	0.8	0.9	-	-	0.89	25	0.04
^{137}Cs	S-3/1	370	380	27	38	0.97	25	0.045
	S-3/4	283	295	65	92	0.96	30	0.034
^{85}Sr	S-3/1	265	270	2.5	3	0.98	50	0.019
	S-3/4	295	310	1.9	2.5	0.95	13	0.078
^{152,154}Eu	S-3/1	>800	>800	126	142	1.0	30	0.035
	S-3/4	>800	>800	15	22	1.0	30	0.036

2.1.2.3. Transport and sorption of ^{85}Sr and ^{125}I in crushed crystalline rocks

The crystalline rocks, synthetic groundwater, equipments and working procedure were in the study presented [21] identical with our previous experiments cited in subchapter 2.1.2.1. The solutions of ^{85}Sr^{2+} and ^{125}I^{-} of several tens of kBq activity of high radiochemical and radionuclidic purity (min. 99%) were applied individually into the stream of SGW in the form of a short pulse of about 0.1 cm^3 volume. The SGW contained carriers as Sr(NO$_3$)$_2$ and NaI in 10^{-6} mol/dm^3 concentration. The flow rate, i.e., the seepage velocity of the SGW or acids mixture simulating acid rain (subchapter 2.1.2.1) through each column was kept on the value 0.18 ± 0.05 cm/min. Pore-water volume was 12.5 ± 0.2 cm^3.

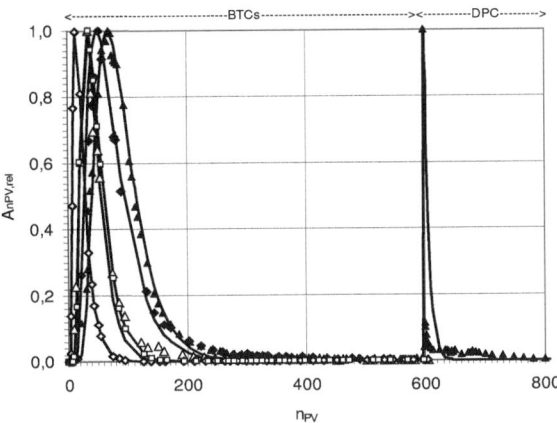

Fig. 8 Experimental (symbols) and theoretical (solid lines) BTCs of ^{85}Sr sorption on diorite-I (▲),diorite-II (♦), gabbro (Δ), granite (◊) and tonalite (□), and DPC of ^{85}Sr desorption fromdiorite-I (▲) (From Palágyi Š et al. (2010) J Radioanal Nucl Chem 283:629. With permission)

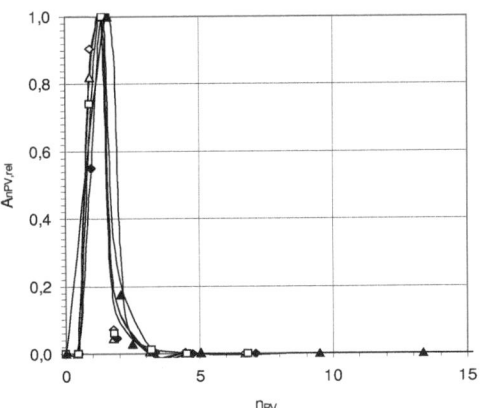

Fig. 9 Experimental (symbols) and theoretical (solid lines) BTCs of ^{125}I sorption on diorite-I (▲), diorite-II (♦), gabbro (Δ), granite (◊) and tonalite (□) (From Palágyi Š et al. (2010) J Radioanal Nucl Chem 283:629. With permission)

Table 3 The most important transport parameters of the sorption of ^{85}Sr and ^{125}I in crushed rocks (From Palágyi Š et al. (2010) J Radioanal Nucl Chem 283:629. With permission)

Crushed rocks	Retardation coefficient		Peak position coefficient	Peclet number	Distribution coefficient [cm³/g]	Peak height correction factor	Hydro-dynamic dispersion coefficient [cm²/min]	WSOS /DF
	$R_{exp,s}$	$R_{th,s}$	k_p	Pe_s	$K_{exp,d,s}$	k_h	$D_{d,s}$	
				^{85}Sr				
Diorite I	69	72	1.04 ± 0.01	8.92 ± 0.02	27	0.86	0.15	4.38
Diorite II	53	65	1.23 ± 0.01	7.65 ± 0.03	19	0.80	0.17	16.60
Gabbro	35	37	1.05 ± 0.01	6.11 ± 0.03	12	0.71	0.21	11.20
Granite	33	37	1.10 ± 0.01	6.0 ± 0.03	12	0.71	0.22	11.60
Tonalite	42	43	1.04 ± 0.01	9.7 ± 0.04	16	0.89	0.13	9.73
				^{125}I				
Diorite I	1.6	1.3	0.84 ± 0.01	38.2 ± 0.1	0.24	1.20	0.03	11.4
Diorite II	1.4	1.2	0.84 ± 0.01	53.1 ± 0.3	0.14	1.24	0.02	46.6
Gabbro	1.3	1.0	0.79 ± 0.01	43.5 ± 0.3	0.11	0.90	0.03	22.4
Granite	1.3	1.0	0.81 ± 0.01	39.9 ± 0.2	0.11	1.06	0.03	22.6
Tonalite	1.4	1.1	0.80 ± 0.01	60.2 ± 0.7	0.15	0.95	0.02	17.5

The experimental BTCs for ^{85}Sr^{2+} and ^{125}I$^-$ were fitted by theoretical BTCs and presented in Figs. 8 and 9. The evaluation of the BTCs was made on the basis of peak maximum activity using Eq. 14. Theoretical BTCs, obtained by means of above mentioned non-linear regression method and the measured points agree well mainly in the peak areas, from which the basic transport parameters are derived, namely the R_{exp} ($\equiv n_{PVmax}$).

The activity balance made for each rock has shown little differences between experimental and theoretical points of the respective BTCs, mainly in the ascending part. The *WSOS/DF* quantity reflects the agreement between calculated and experimental values of BTCs in case of ^{85}Sr^2. Displacement experiments with acids mixture proved that ^{85}Sr can be quantitatively desorbed from crushed rocks, which is confirmed by high value of the hysteresis coefficient, $K_h = 0.99$ ($K_h = (1 - K_{d,des}/K_{d,s})$. The K_h is generally regarded as a measure of sorption irreversibility and can be defined differently [29]. The results show that the degree of irreversibility is very low.

The values of *WSOS/DF* quantity for ^{125}I listed in Table 3 are on average higher than the generally acceptable: $0.1 \leq WSOS/DF \leq 20$ – see above. First of all, it can be in consequence of the relatively small number of experimental points obtained - as a result of high transport rates - in the course of a given experiment. From the results it can be seen that the transport of [125]I$^-$, i.e., iodine in the chemical form of iodide, is very fast as practically no sorption of I$^-$ takes place, due to the negatively charged sorption centers of the crushed rocks [30].

2.1.2.4. Transport of [125]I$^-$, [137]Cs$^+$ and [85]Sr^{2+} in granitic rock and soil

Three types of rocks from the locality of the Cavernous Gas Reservoir, near Příbram (Czech Republic) were taken in the study: diorite-I, diorite-II and tonalite [22]. A homogenized fraction of $0.25 - 0.8$ mm was used. Samples of soil from the vicinity of High Level Waste Storage Facility in the area of Nuclear Research Institute Řež in Husinec-Řež (Czech Republic) were taken from 5-20 cm depth (surface samples). The samples were homogenized. The rock samples were packed into medical plastic syringes, serving as columns of 2.1 cm in diameter and 8.8 cm in length. The pore water volume was 11.2 ± 1.3 cm^3. Soil samples were placed in dual column as in subchapter 2.1.2.2. Pore-water volume was 30.4 ± 3.8 cm^3. Synthetic groundwater was the same as in subchapter 2.1.2.1, and contains the respective carriers in 10^{-6} mol/dm^3 concentrations. For delivering of SGW at a constant flow rate (0.3 cm^3/min in rock and 0.1 cm^3/min in soil), the forced flow was applied using a multi-head peristaltic pump.

Experimental and theoretical (fitted) breakthrough curves of [137]Cs$^+$ and [85]Sr^{2+} transport in selected crushed rock of diorite-I, based on both linear and non-linear sorption models, are presented in Figs. 10 and 11. The iodine was omitted as it did not sorb at all. The rest of two studied granitites (diorite-II and tonalite), were omitted in these figures, as their BTCs are close to each other and is hardly distinguishable. Resulting parameters are collected in Table 4. Generally, the soil has a greater sorption capacity, therefore the higher retardation capability than for crushed rock samples, was observed.

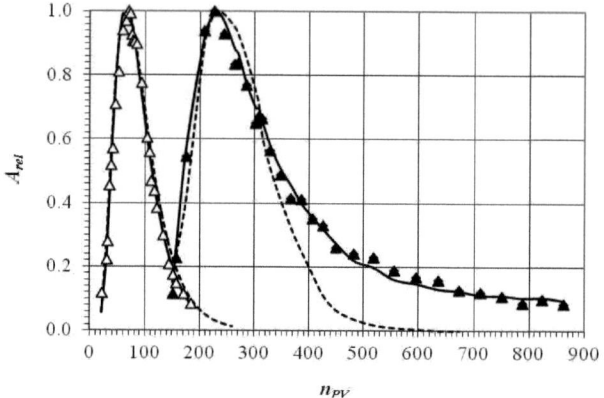

Fig. 10 Transport of $^{85}Sr^{2+}$ (Δ) and $^{137}Cs^{+}$ ($\blacktriangle$) - experimental values in crushed diorite I (symbols) and their fitting using Eq. (14). Linear sorption isotherm approach: dashed line (---) and non-linear sorption isotherm approach: solid line (—) (From Palágyi Š and Štamberg K (2010) J Radioanal Nucl Chem 286:309. With permission)

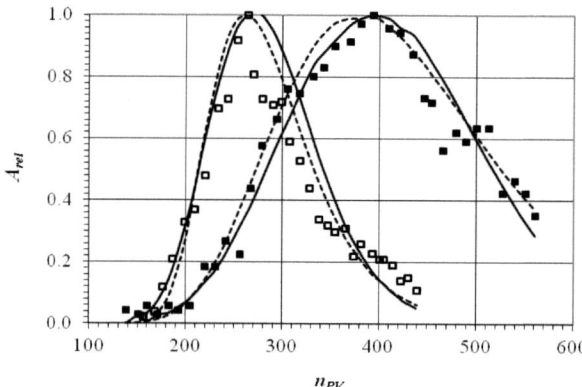

Fig. 11 Experimental values of $^{85}Sr^{2+}$ ($\square$) and $^{137}Cs^{+}$ ($\blacksquare$) transport in surface soil and their fitting using Eq. (14). Legends see Fig. 2 (From Palágyi Š and Štamberg K (2010) J Radioanal Nucl Chem 286:309. With permission)

Table 4 The most important transport parameters of ^{137}Cs$^+$ and ^{85}Sr^{2+} sorption in crusheddiorite I (Dior I) and surface soil (Palágyi Š and Štamberg K (2010) J Radioanal Nucl Chem 286:309. With permission)

Linear sorption isotherm approach										
Radio-nuclide	Sample	R_{exp} -	$K_{d,exp}$ g cm^{-3}	R_{cal} -	$K_{d,cal}$ g cm^{-3}	k_p -	k_h -	Pe -	D_d cm^2 min^{-1}	WSOS /DF
^{137}Cs$^+$	Dior I	227	84.2	232	86.0	1.02	1.22	22.4	0.06	25
	Soil	370	138.1	392	143.8	1.06	1.44	25.6	0.04	24
^{85}Sr^{2+}	Dior I	69	25.3	72	26.4	1.04	0.86	8.9	0.16	4.4
	Soil	262	97.8	275	100.8	1.05	1.98	49.6	0.02	22
Non-linear sorption isotherm approach										
Radio-nuclide	Sample	R_{cal} -	$K_{d,exp}$ g cm^{-3}	k_F m^3 kg^{-1}	n_F -	k_p -	k_h -	Pe -	D_d cm^2 min^{-1}	WSOS /DF
^{137}Cs$^+$	Dior I	258 - 565	84.2	22.9	0.91	1.00	0.27	23.0	0.10	2.8
	Soil	223 - 412	138.1	294	1.03	0.94	1.16	31.9	0.04	18
^{85}Sr^{2+}	Dior I	72 - 74	25.3	24.7	0.99	1.00	0.84	8.9	0.16	3.3
	Soil	269 - 307	97.8	91.2	0.98	1.00	1.97	52.4	0.03	22

For the evaluation of experimental data the Newton-Raphson multi dimensional method of non-linear regression was used [18], by means of which the data are fitted using a regression function based on Eqs. 7 and 14. From Fig. 10 we see that, first of all, the great difference exists between fitting of ^{137}Cs$^+$ and ^{85}Sr^{2+} experimental points, so it is evident that in a case of strontium the linear isotherm characterizes its sorption. It does not hold for cesium, where in addition the linear sorption isotherm is more suitable in the top part of the curve, and non-linear one in the low part.

From the comparison of Fig. 10 with Fig. 11, especially in a case of ^{137}Cs$^+$, it is seen that Eq. (14), based on correction coefficients, gives considerably better goodness-of-fit than the application of Eq. (11). Some experimental and calculated parameters regarding the experiments presented in Figs. 10 and 11 can be found in Table 4. It is evident, especially according to the *WSOS/DF* values, that the non-linear isotherm sorption approach is better than the linear one - only with one exception (see sorption of ^{85}Sr^{2+} on soil) - in case of Sr it seems that the linear approach is quite satisfactory. It is possible to judge the difference between both approaches by comparison of equilibrium parameters $K_{d,exp}$ - $K_{d,cal}$ - k_F - n_F, the values of which are also collected in Table 4. We can see that n_F moves more toward 1, k_F approaches $K_{d,exp}$ or $K_{d,cal}$, and the difference between linear and non-linear approaches is disappearing.

2.2. Step (continuous) tracer inlet of radioactive contaminants

2.2.1. Radioactive contaminant transport modeling

Transport of a contaminant under equilibrium dynamic conditions between mobile (liquid) and stationary (solid) phase can be described by Eq. (1), i.e., with the well-known 1-D ADE. By integration under suitable boundary conditions [$C(0,t)$, $x(0,L)$] with consecutive changes of initial conditions [C (or A) = 0 at $x = 0$ and $t_0 = 0$, and C_t (or A_t) = C_0 (or A_0) at $x = L$], and rearranging, the following equation describing the sorption breakthrough curve can be obtained for C_{rel} or A_{rel} [27,31-33]:

$$C_{rel} = \frac{C}{C_o} \equiv A_{rel} = \frac{A}{A_o} = \frac{1}{2}\operatorname{erfc}\left(\frac{R_{cal} - n_{PV}}{2\sqrt{\dfrac{R_{cal} n_{PV}}{Pe}}} \right) . \tag{15}$$

The constant initial contaminant concentration and constant coefficients are necessary preconditions in this case. Again, two approaches, namely based on the linear and non-linear equilibrium isotherm, can be used. Because the sorption and desorption breakthrough curves are theoretically mirror-symmetrical, there are two basic corresponding equations, namely Eqs. (16) and (17) at hand:

$$A_{relS} = \frac{A_S}{A_0} = \frac{1}{2}\operatorname{erfc}\left(\frac{R_{Scal} - n_{PV_S}}{2\sqrt{\dfrac{R_{Scal} n_{PV_S}}{Pe}}} \right) \tag{16}$$

$$A_{relD} = \frac{A_D}{A_0} = 1 - \frac{1}{2}\operatorname{erfc}\left(\frac{R_{Dcal} - n_{PV_D}}{2\sqrt{\dfrac{R_{Dcal} n_{PV_D}}{Pe}}} \right) \tag{17}$$

S and D are indexes for sorption and desorption, respectively. By analogy with the pulse tracer inlet modeling, it holds for the points of inflection (see index i) of both curves: $R_{Sexp} = n_{PVSi}$ and $R_{Dexp} = n_{PVDi}$. These relations can be

24

used for the estimation of corresponding retardation coefficients. Of course, more exact values, R_{Scal} and R_{Dcal}, are obtained by non-linear regression procedure of experimental data.

2.2.2. Evaluation of step tracer inlet experimental data

2.2.2.1. Determination of ^{137}Cs and [85]Sr transport parameters in fucoidic sand

The borehole samples of fucoidic sandstones, studied in work [23], were taken from two localities of the former uranium underground (in-situ) leaching area in Stráž pod Ralskem (Czech Republic). Undisturbed cores of 5 cm in diameter and 10 cm height (ca 360 – 380 g) were embedded in the Eprosin type cured epoxide resin column. The pore water volume was between 50 – 58 cm^3, depending on the column. The core sample and a part of the experimental setup are presented in Fig. 12. Results of the chemical bulk analysis and some important physical parameters of sands along with a composition of groundwater, mainly of ammonium sulfate type, originating from Cenomanian aquifer are given in [23]. Radionuclides and carrier contents of groundwater were the same as given in previous subchapters. The sorption experiments were carried out separately for ^{137}Cs or [85]Sr. The active groundwater from a suitable large beaker (of several litres) was introduced at the bottom of columns (against gravity), by the use of multi-head peristaltic pump, with flow-rate of 4.4 ± 0.2 mL h^{-1} (0.78 ± 0.01 cm h^{-1} water seepage velocity). After the input activity at the column outlet had been attained, the beaker was filled with inactive groundwater, which was further used for desorption of a given radionuclide at the same flow-rate. Breakthrough curves (Figs. 13 and 14) were calculated according Eqs. 16 and 17, with the use of own computational code (PNLRPa11.fm constructed in FAMULUS software product [19]).

Fig. 12 Core sample of the fucoidic sand and a part of the experimental setup. (From Palágyi Š and Štamberg K (2011) Cent Eur J Chem 9:798. With permission)

A fairly good agreement has been found between experimental and theoretical sorption/desorption breakthrough curves and transport parameters of Cs^+ and Sr^{2+} in systems consisted of fucoidic sand-cores and groundwater, especially if the non-linear isotherm approach is used. On the other hand, sorption of both radionuclides on fucoidic samples is unambiguously characterized with linear isotherms (compare the values of $(K_{dS})_{theor} - (K_{dS})_{exp} - k_S - n_S$ in Tables 5 and 6), and therefore it can be evaluated satisfactorily with linear isotherm approach. From this point of view, the similarity between $(K_{dS})_{theor}$ and $(K_{dS})_{exp}$ comes up to expectations. The resulted values show that the sorption capacity of the fucoidic samples studied for cesium is about 3-times higher than that for strontium at 10^{-6} mol. dm^3 carrier concentration. However, desorption with groundwater originating from Cenomanian aquifer is more effective in the case of cesium than that of strontium probably due to the relatively high concentration of Ca^{2+} in CGW and to the above mentioned ion-exchange mechanism.

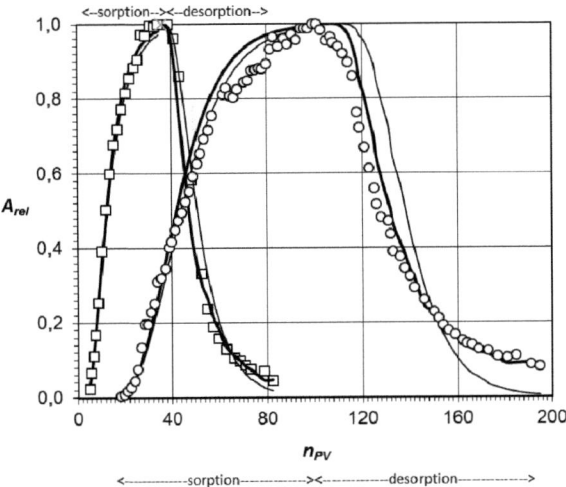

Fig. 13 Fitting of the sorption and desorption BTCs of ^{137}Cs in samples No. 48562Cs (□) and 48118Cs (○) (experimental values: symbols, and theoretical: thin curves – linear isotherm approach, thick curves – non-linear isotherm approach). (From Palágyi Š and Štamberg K (2011) Cent Eur J Chem 9:798. With permission)

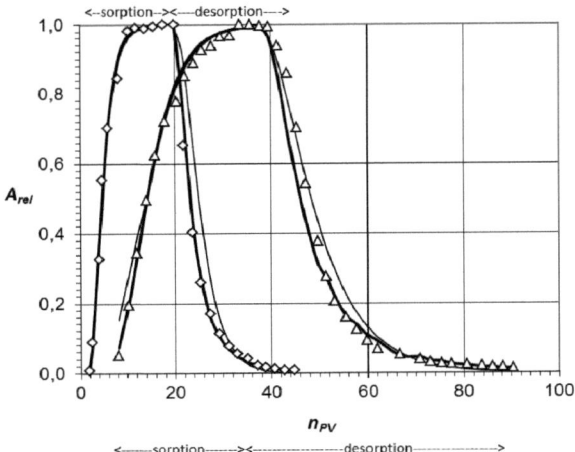

Fig. 14 Fitting of the sorption and desorption BTCs of ^{85}Sr in No. 48562Sr (◊) and 48118Sr (Δ) (experimental values: symbols and theoretical: thin curves – linear isotherm approach, thick curves – non-linear isotherm approach). (From Palágyi Š and Štamberg K (2011) Cent Eur J Chem 9:798. With permission)

Table 5 Important transport parameters of the sorption and desorption of ^{137}Cs and ^{85}Sr in fucoidic sand columns – linear isotherm approach. (From Palágyi Š and Štamberg K (2011) Cent Eur J Chem 9:798. With permission)

Radio-nuclide	Column	$(K_{dS})_{theor}$ or $(K_{dD})_{theor}$ mL g^{-1}	$(K_{dS})_{exp}$ or $(K_{dD})_{exp}$ mL g^{-1}	Pe	D_d cm^2 h^{-1}	$(R_S)_{theor}$ or $(R_D)_{theor}$	$(R_S)_{exp}$ or $(R_D)_{exp}$	WSOS /DF
				Sorption				
^{137}Cs	48118	6.41±0.03	7.25	15.4±0.2	0.45	44.1	44.7	12.3
	48562	1.67±0.01	1.67	6.8±0,1	1.20	13.0	12.5	5.27
^{85}Sr	48118	1.90±0.01	1.91	7.4±0.1	1.11	14.8	14.0	16.6
	48562	0.58±0.01	0.57	9.5± 0.1	0.87	4.72	4.5	20.9
				Desorption				
^{137}Cs	48118	5.61±0.04	4.10	15.4±0.2	0.45	38.7	25.7	12.3
^{137}Cs	48562	2.11±0.11	1.49	6.8±0.1	1.20	16.1	11.2	5.27
^{85}Sr	48118	1.90±0.01	1.68	7.4±0.1	1.11	14.9	12.4	16.6
^{85}Sr	48562	1.00± 0.01	0.64?	9.5±0.1	0.87	7.83	4.9 ?	20.9

Table 6 Important transport parameters of the sorption and desorption of ^{137}Cs and ^{85}Sr in fucoidic sand columns – non-linear isotherm approach. (From Palágyi Š and Štamberg K (2011) Cent Eur J Chem 9:798. With permission)

Radio-nuclide	Sample (column)	k_S or k_D mL g^{-1}	n_S or n_D	Pe	D_d cm^2 h^{-1}	$(R_S)_{theor}$ or $(R_D)_{theor}$	$(R_S)_{exper}$ or $(R_D)_{exper}$	WSOS /DF
				Sorption				
^{137}Cs	48118	5.24±0.12	0.990±0.001	16.1±0.1	0.43	41.8-44.1	44.7	2.16
	48562	1.72±0.10	1.000±0.003	7.9±0.1	1.03	12.5-12.7	12.5	0.97
^{85}Sr	48118	1.80±0.16	1.000±0.006	13.5±0.2	0.61	14.2	14.0	1.67
	48562	0.78±0.05	1.020±0.004	11.4±0.1	0.72	4.49-4.83	4.5	1.34
				Desorption				
^{137}Cs	48118	0.17±0.01	0.750±0.002	16.1±0.1	0.43	27.9-49.2	25.7	2.16
	48562	0.13±0.01	0.800±0.004	7.94±0.08	1.03	12.0-21.0	11.2	0.97
^{85}Sr	48118	0.14±0.01	0.810±0.002	13.5±0.2	0.61	11.7- 24.9	12.4	1.67
	48562	0.09±0.01	0.840±0.002	11.4±0.1	0.72	5.45-10.6	4.9 ?	1.34

2.2.2.2. Effect of grain size on the transport parameters of ^{137}Cs in pure granite and fracture infill

Two types of crystalline rocks were investigated in work [24], as follows: pure granite coded as PDM1-1 and fracture filling materials (clays) coded as PDM1-2. Rocks were sampled from PDM1 borehole, the samples PDM1-1 from 97.5-98.7 m depth and PDM1-2 from 89.7−90.0 m depth. Each rock was crushed and sieved to $0.063−0.125$, $0.125−0.630$, $0.63−0.80$ and $0.80−1.25$

mm fractions. The fractions were placed into 5 cm³ suitably adapted plastic columns (medical syringes) of 1.3 cm inner diameter and 5.4 cm height. The pore volumes were between 2.8 and 3.2 cm³. Aqueous solution of $^{137}Cs^+$ tracer with a 10^{-6} mol/dm³ carrier concentration was added into a suitable reservoir of the synthetic groundwater (SGW), the composition of which, pH and ionic strength were given in [24]. The experimental setup of the experiments is depicted in Fig. 1. The constant rate-flow was ca 0.05 cm³/min. Tracer-free SGW without the added carrier was used for the elution (desorption process).

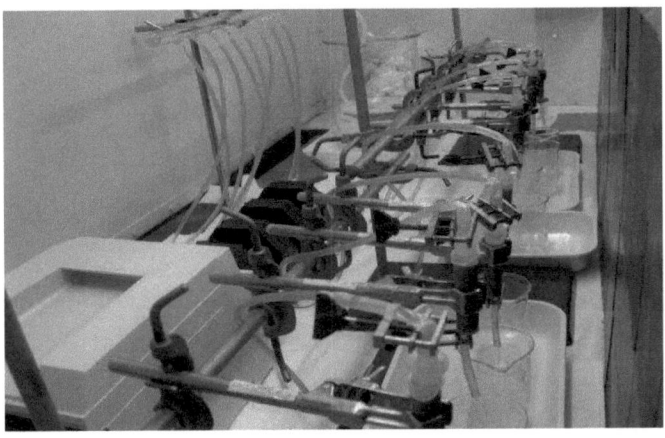

Fig. 15 Setup used in the dynamic experiments with crushed granite and fillings (From Štamberg K and Palágyi Š (2011) J Radioanal Nucl Chem 293:127. With permission)

The experimental breakthrough data of $^{137}Cs^+$ sorption obtained with crushed granites and their filling materials, as well as the results of the fitting of these experimental data with theoretical BTCs calculated by linear and non-linear isotherm approach are presented in Figs. 16 and 17, respectively. It is seen that a satisfactorily fitting of sorption BTCs can be obtained using both model approaches. Fitting of experimental data with non-linear isotherm approach model is somewhat more effective as is documented in Tables 7 and 8 by WSOS/DF values.

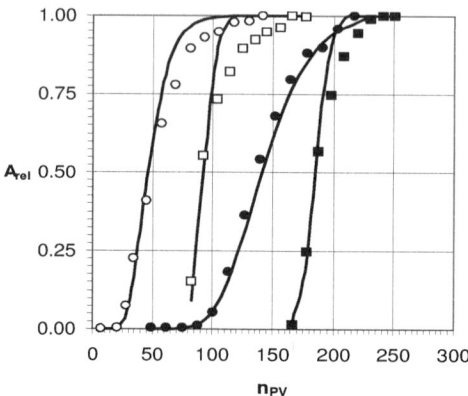

Fig. 16 Experimental (symbols) and theoretical (solid lines) BTCs of $^{137}Cs^+$ sorption in crushed granite (empty symbols) and filling material (full symbols) of 0.063-0.125 mm ($\square$, $\blacksquare$) and 0.63 – 0.80 mm grain size ($\circ$, $\bullet$). Theoretical BTCs were calculated using linear isotherm model (From Štamberg K and Palágyi Š (2011) J Radioanal Nucl Chem 293:127. With permission)

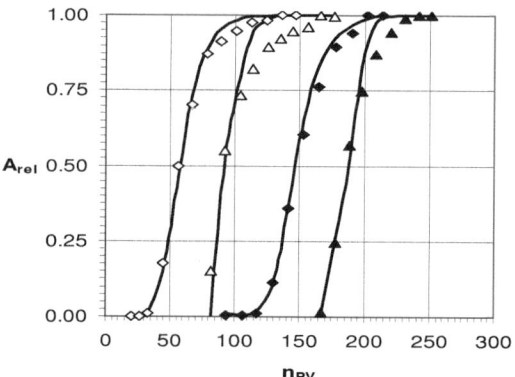

Fig.17 Experimental (symbols) and theoretical (solid lines) BTCs of $^{137}Cs^+$ desorption in crushed granite (empty symbols) and fillings (full symbols) of 0.063-0.125 mm ($\square$, $\blacksquare$) and 0.63-0.80 mm grain size ($\lozenge$,$\blacklozenge$). Theoretical BTCs were calculated using non-linear isotherm approach model (From Štamberg K and Palágyi Š (2011) J Radioanal Nucl Chem 293:127. With permission)

In accordance with the values of WSOS/DF, non-linear isotherm approach model generally provides better agreement with experimental data than linear isotherm approach one particularly in cases, when parameters of $n_{S,D}$ differ from 1 ($n_{S,D} \neq 1$). If $n_{S,D} \approx 1$, then $k_{S,D} \approx (K_{dS,D})_{theor}$ and linear isotherm approach holds.

Table 7 The important transport parameters of $^{137}Cs^+$ in crushed granite (PDM1-1a,b,c,d) and filling (PDM1-2a,b,c,d) material – calculated using linear isotherm approach model (From Štamberg K and Palágyi Š (2011) J Radioanal Nucl Chem 293:127. With permission)

Column No.	$(R_{S,D})_{exp}$ -	$(R_{S,D})_{theor}$ -	$(K_{d,S,D})_{exp}$ cm^3/g	$(K_{d,S,D})_{theor}$ cm^3/g	Pe -	$D_d \times 10^{-3}$ cm^2/min	$WSOS/DF$ -
Sorption							
PDM1-1a	91.4	92.7	34.0	34.5±0.2	253±15	1.568	1.89
PDM1-1b	81.7	76.6	24.2	22.7±0.1	24.7±0.1	17.205	8.46
PDM1-1c	56.0	60.0	16.4	17.6±0.1	26.4±0.1	16.438	1.58
PDM1-1d	50.1	46.4	14.7	13.6±0.1	21.5±0.6	21.162	11.7
PDM1-2a	185.9	185.0	69.0	68.5±0.1	757±8	0.498	1.33
PDM1-2b	205.5	199.0	61.1	59.1±0.3	102±2	3.997	34.3
PDM1-2c	147.9	167.0	44.8	50.4±0.3	77±2	5.688	18.2
PDM1-2d	137.0	141.0	41.3	42.6±0.1	41.4±0.5	11.118	19.1
Desorption							
PDM1-1a	92.0	65.0	34.1	24.1±0.8	2.6±0.2	152.58	5.70
PDM1-1b	64.0	67.9	18.9	20.1±0.4	3. 9±0.2	107.806	1.71
PDM1-1c	38.4	35.2	10.7	10.2±0.2	1.5±0.1	289.31	1.12
PDM1-1d	40.0	37.7	19.7	11.0±0.1	2.2±0.1	211.286	1.11
PDM1-2a	235.0	181.0	87.3	67.2±1.2	10.1±0.7	37.325	5.40
PDM1-2b	211.9	200,0	63.5	59.5±0.7	9.0±0.4	45.33	1.94
PDM1-2c	141.2	125.0	42.7	37.8±0.4	3.7±0.1	118.41	1.36
PDM1-2d	141.2	132.0	42.6	39.8±0.5	8.5±0.4	53.77	1.76

The experimental data of $^{137}Cs^+$ desorption from the above mentioned granites and filling materials are presented in Fig. 18. The fitted curves for both linear and non-linear isotherm model are of a tight vicinity, so only breakthrough curves (BTC_D) obtained by non-linear isotherm model (according to Eq. 17) are depicted here. As is seen in Tables 7 and 8 that in the case of desorption the fitting by linear isotherm approach model is acceptable because the WSOS/DF values fulfill the above mentioned condition (0.1 ≤ WSOS/DF ≤ 20) well. But, on the other hand, the Freundlich equation parameters presented in Table 9 (see $n_{S,D}$ values), clearly show the deviation from linearity.

Table 8 The most important transport parameters of $^{137}Cs^+$ in crushed granite (PDM1-1a,b,c,d) and filling (PDM1-2a,b,c,d) material – calculated using non-linear isotherm (From Štamberg K and Palágyi Š (2011) J Radioanal Nucl Chem 293:127. With permission)

Column No.	$k_{S,D}$ cm³/g	$n_{S,D}$ -	$(R_{S,D})_{theor}$ -	Pe -	$D_d \times 10^{-3}$ cm²/min	WSOS/DF -
			Sorption			
PDM1-1a	87.3±5.6	1.060	73-104	264±10	1.50	0.68
PDM1-1b	265±1	1.010	71-81	25.2±0.1	17.00	7.94
PDM1-1c	14.5±0.2	0.990	56-62	26.1±0.1	16.70	1.26
PDM1-1d	2.29±0.05	0.873	40-83	4.59±0.01	99.10	1.01
PDM1-2a	76.0±0.7	1.010	183-189	764±5	0.50	0.67
PDM1-2b	19.8±0.1	0.920	188-474	13.2±0.1	30.90	24.7
PDM1-2c	26.6±0,2	0.965	139-167	70.7±0.2	6.20	2.31
PDM1-2d	15.8±0.4	0.929	131-212	13.6±0.1	33.80	9.04
			Desorption			
PDM1-1a	1520±618	1.31	28-69	1.0±0.1	396.0	4.89
PDM1-1b	19200±8890	1.51	24-80	1.04±0.08	411.92	4.15
PDM1-1c	41.1±7.9	1.11	25-35	1.10±0.08	396.24	0.84
PDM1-1d	4.56±0.66	0.93	38-50	3.17±0.21	143.49	1.06
PDM1-2a	482±193	1.14	183-204	6.18±0.96	61.81	4.93
PDM1-2b	10700±3650	1.38	170-251	2.38±0.18	171.38	1.26
PDM1-2c	11.7±3.3	0.91	123-146	6.03±0.76	72.69	1.29
PDM1-2d	10100±3420	1.41	65-164	2.09±0.14	219.94	1.19

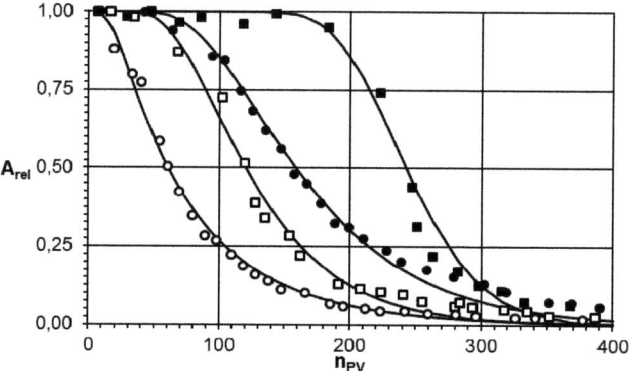

Fig. 18 Experimental (symbols) and theoretical (solid lines) BTCs of $^{137}Cs^+$ desorption in crushed granite (empty symbols) and fillings (full symbols) of 0.063-0.125 mm (□, ■) and 0.63-0.80 mm grain size (◊,♦). Theoretical BTCs were calculated using non-linear isotherm (From Štamberg K and Palágyi Š (2011) J Radioanal Nucl Chem 293:127. With permission)

It follows from our previous works and is also clear from the theoretical part of this paper that on the contrary to linear model, the value of the retardation coefficient in non-linear model is not constant during the sorption in a given experimental arrangement, but it changes with n_{PV} as can be seen in Fig. 19. Therefore in Table 8 not one single value is presented, but the span of the $(R_{S,D})_{theor}$ -values is given for each sample. Desorption $(R_D)_{theor}$ -values have shown much lower differences in the same dependence as sorption retardation coefficients (they are not presented in this figure). The values of $(R_{S,D})_{theor}$ (see Table 7) calculated using linear isotherm approach model equal approximately to $(R_{S,D})_{exp}(= n_{PVS,Di})$ estimated from point of inflection of corresponding breakthrough curves. It is evident that the agreement corresponds with WSOS/DF values, because the smaller is the WSOS/DF, the better is the agreement.

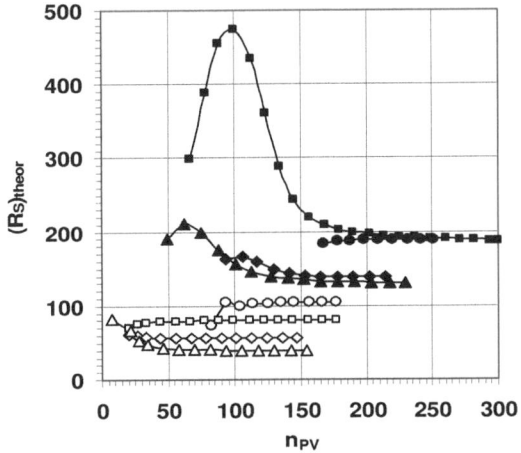

Fig. 19 Dependence of sorption $(R_S)_{theor}$ -values on n_{PV} of $^{137}Cs^+$ transport in granite (empty symbols) and fillings (full symbols) of $0.063 - 0.125$ mm $(\circ, \bullet)$, $0.125 - 0.63$ mm $(\square, \blacksquare)$, $0.63 - 0.80$ $(\Diamond, \blacklozenge)$ and $0.80 - 1.25$ mm $(\triangle, \blacktriangle)$ grain size. The $(R_S)_{theor}$ values were calculated using non-linear isotherm approach model (From Štamberg K and Palágyi Š (2011) J Radioanal Nucl Chem 293:127. With permission)

The effect of the average grain size on the sorption and desorption $(R_{S,D})_{theor}$ – values is demonstrated in Fig. 20. The dependences show that the grain size has a considerably effect on the retardation of $^{137}Cs^+$, as well as its distribution coefficients. Both coefficients generally increase with decreasing grain size. The reason lies mainly in the increase of accessible active sorption centers due to increase in specific surface of the rock materials, as well as owing to uncovering other functional groups positively influencing sorption of Cs^+ ions. This finding is in a good accordance with results obtained previously by several authors [see them in Ref. 24]. The increase is marked especially in sorption on pure granite, where K_d for 1.025 mm average grain size is approx. 14 cm^3/g, but for 0.094 mm average grain size is as much as 34 cm^3/g, i.e., about 2.3-times higher. In respective filling materials this increase is somewhat slighter of about 1.7-times only.

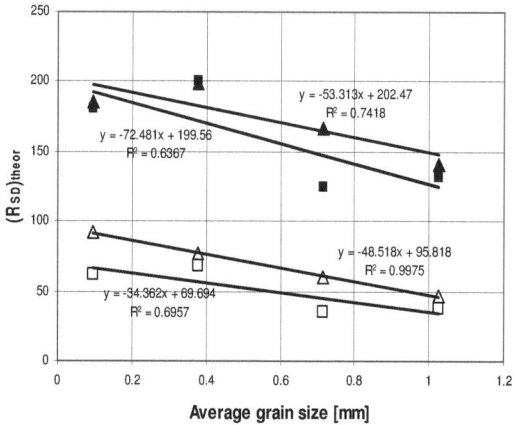

Fig. 20. Effect of the average grain size on the desorption ($\square$, $\blacksquare$) and sorption ($\triangle$, $\blacktriangle$) $(R_{S,D})_{theor}$ -values of $^{137}Cs^+$ transport in granite (empty symbols) and fillings (full symbols). The $(R_{S,D})_{theor}$ -values were calculated using linear isotherm approach model

2.2.2.3. Effect of grain size on the $^{85}Sr^{2+}$ transport parameters in pure granite and fracture infill

Two types of crystalline rocks were investigated [25]: pure granite, coded as PDM1-1, and fracture infill material (formed namely by clay minerals), coded as PDM1-2. Both types of rocks were sampled from the PDM1 borehole, as given in subchapter 2.2.2.2. The borehole samples are depicted in Fig. 21. The X-ray phase analysis and the content of corresponding mineralogical phases in detail are given in our paper [24]. Each rock was crushed and sieved to $0.063 - 0.125$ mm, $0.125 - 0.630$ mm, $0.63 - 0.80$ mm and $0.80 - 1.25$ mm fractions. The average grain sizes (AGS) were as follows 0.094, 0.377, 0.715 and 1.025 mm, respectively. The fractions were placed into 5 cm^3 PET columns of 1.3 cm inner diameter and 5.4 cm height. The pore volumes were 3.0 ± 0.2 cm^3. The basic hydraulic parameters of column filled with crushed rocks were calculated using experimentally determined values of mass and volume of the column filled with samples of a given grain size. Aqueous solution of $^{85}Sr^{2+}$ tracer (1.10^{-13} mol/dm^3, about 0.1 kBq/cm^3) with a 10^{-6} mol/dm^3 $Sr(NO_3)_2$ concentration was added into a suitable reservoir of the synthetic groundwater. The composition of SGW was, in mg/dm^3, as follows: Na^+ - 10.64, K^+ - 1.80, Ca^{2+} - 27.00, Mg^{2+} - 6.40, Sr^{2+} - 0.02, Cl^- - 42.40, SO_4^{2-} - 27.70, NO_3^- - 6.30, HCO_3^- - 30.41, F^- - 0.20, pH = 7.71). This solution was used for sorption of $^{85}Sr^{2+}$ in solid crushed rock material. Tracer-free SGW without the added carrier was used for elution (desorption) of studied species from the layer of crushed rock. The experimental setup was the same as depicted in [24]. The initial activity of the aqueous phase and its flow-rate were constant during the sorption process. The rate-flow was between 0.05 $- 0.06$ cm^3/min.

Fig. 21 Borehole samples: pure granite (left) and fracture infill (right) (From Palágyi Š et al. (2013) J Radioanal Nucl Chem 297:33. With permission)

The experimental BTC data of $^{85}Sr^{2+}$ sorption and desorption obtained with crushed granites and the fracture infill, as well as the results of the fitting of these data with theoretical BTCs, calculated by linear and non-linear isotherm approach models, are presented in Figs. 22 and 23. The BTCs of sorption and desorption are almost symmetric, as has been assumed theoretically.

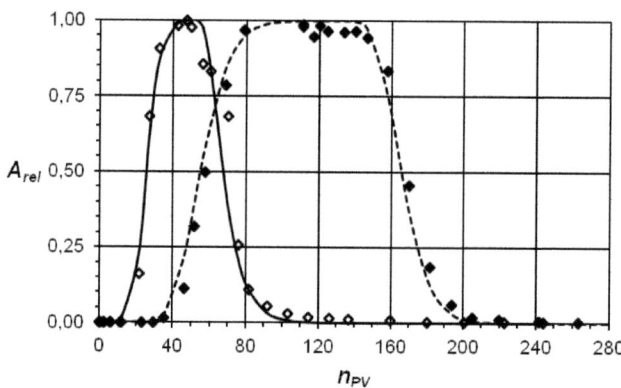

Fig. 22 Experimental (symbols) and theoretical (curves) BTC of $^{85}Sr^{2+}$ sorption (ascending curves) and desorption (descending curves) in granite ($\Diamond$, —) and fracture infill ($\blacklozenge$, ----), both of 0.063-0.125 mm grain size. Theoretical curves were calculated using linear isotherm (From Palágyi Š et al. (2013) J Radioanal Nucl Chem 297:33. With permission)

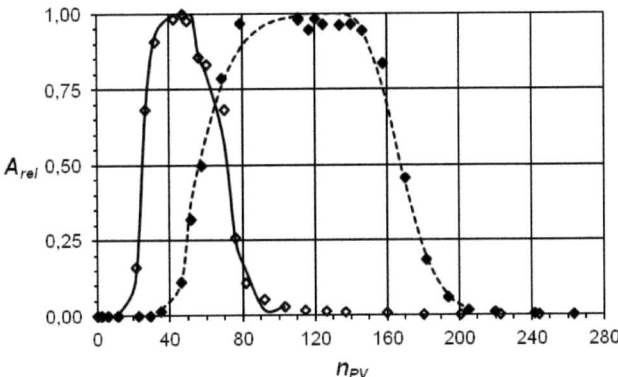

Fig. 23 Experimental (symbols) and theoretical (curves) BTC of $^{85}Sr^{2+}$ sorption (ascending curves) and desorption (descending curves) in granite ($\lozenge$, —) and fracture infill ($\blacklozenge$, ----) of 0.063 – 0.125 mm grain size. Theoretical curves were calculated using non-linear isotherm (From Palágyi Š et al. (2013) J Radioanal Nucl Chem 297:33. With permission)

This fact confirms previous perceptions that the sorption of $^{85}Sr^{2+}$ is reversible under dynamic conditions [21,22], and that both sorption and desorption processes are fast. Result of experimental data fitting with non-linear isotherm approach is somewhat better than that with linear isotherm one, as is documented in Tables 3 and 4 – compare the values of fitting criterion *WSOS/DF*; particularly in cases when Freundlich equation parameters type of $n_{S,D}$ differ from 1. If $n_{S,D} \approx 1$, then $k_{S,D} \approx (K_{dS,D})_{theor}$ and linear isotherm approach holds.

Similarly, as in previous cases, a defined amounts from aqueous phase, in this case 3 cm^3 volume samples were taken in defined time (t) intervals at the column outlet for gamma activity measurement (A_t), until the activity at the outlet reached the activity at the column inlet (A_0). From the relative activities (A_t/A_0) and the number of pore volumes (n_{PV}) breakthrough curves were constructed (A_t/A_0 vs. n_{PV}). In desorption the sampling of liquid phase was performed until the activity at the column outlet (A_t) attained a value near 0. The breakthrough curves were constructed as A_t/A_{t0} vs. n_{PV}. In this case $A_{t0} = A_0$ is the initial activity at the starting time (t_0) of desorption.

Both breakthrough curves in Figs. 22 and 23, respectively, have shown considerably great difference between sorption of $^{85}Sr^{2+}$ in pure granite and fracture infill as the sorption is considerably higher in the latter. This is also seen in Tables 9 and 10, where these BTCs have been quantified. It is evident that in accordance with sorption capacity, sorption of $^{85}Sr^{2+}$ in fracture infill is about 2.5-times higher in average than for pure granite, independently on the grain size. The sorption capacity for the average grain sizes of 0.094, 0.377, 0.715 and 1.025 mm of pure granite is 3.0, 2.5, 1.6 and 1.2 and fracture infill is 6.8, 6.5, 3.9 and 3.7 mmol/g $\cdot 10^{10}$, respectively.

The study of the $^{85}Sr^{2+}$ sorption/desorption on crushed granite and on the corresponding fissure infill material from synthetic granitic water under dynamic conditions was focused on two problems, namely, (i) the magnitude of sorption capacity of both materials studied, and (ii) the modeling of sorption/desorption (migration/transport) process under column set-up on the laboratory scale.

For the first case, it can be concluded that the results demonstrate the great differences in sorption properties for both materials. The distribution coefficients, sorption capacities and other parameters were found to be approx. 2.5 – 3 times smaller for pure granite than for infill materials. Such observations can play an important role in predicting of contaminant migration through so-called natural host barriers surrounding deep geological repository of radioactive wastes (e.g., used in the course of its performance assessment). The differences were found to be dependent on the grain size too, namely, the smaller is the grain size the higher is the sorption capacity due to the greater specific surface area of crushed material of the smaller grain size. For the second case, the model based on the erfc-function combined with the linear sorption/desorption equilibrium isotherm (K_d-model) and/or with Freundlich equation, generally

38

used for the description of non-linear equilibrium isotherm, proved to be applicable in modeling of such systems.

Table 9 The most important transport and sorption/desorption experimental and theoretical (calculated) parameters of $^{85}Sr^{2+}$ in crushed granite (PDM1-1a, 1b, 1c, 1d) and fracture infill (PDM1-2a, 2b, 2c, 2d) columns. Linear isotherm approach (From Palágyi Š et al. (2013) J Radioanal Nucl Chem 297:33. With permission)

Sample code	$R_{S,exp}$	$R_{D,exp}$	$R_{S,theor}$	$R_{D,theor}$	$K_{dS,theor}$	$K_{dD,theor}$	$Pe_{S,D}$	$D_{dS,D}$	WSOS/DF
	-	-	-	-	cm³/g	cm³/g	-	cm²/min	-
PDM1-1a	24.8	25.2	17.9	19.0	6.39 ± 0.40	6.69 ± 0.15	5.31 ± 0.27	0.080	12.8
PDM1-2a	57.3	57.6	54.8	62.1	20.3 ± 0.12	23.7 ±0.16	191.0 ± 14.5	0.002	18.7
PDM1-1b	24.5	23.4	23.3	22.5	6.46 ± 0.14	6.23 ± 0.04	10.40 ± 0.14	0.042	19.4
PDM1-2b	57.7	63.7	58.3	66.7	16.7 ± 0.06	19.2 ±0.06	25.0 ± 0.27	0.016	3.36
PDM1-1c	12.6	10.1	10.7	8.3	2.91 ±0.14	2.10 ± 0.05	1.75 ± 0.05	0.250	3.93
PDM1-2c	35.1	37.2	34.5	39.3	9.96 ± 0.08	11.6 ± 0.09	89.7 ± 3.98	0.005	10.4
PDM1-1d	11.1	10.0	8.3	8.3	2.30 ± 0.14	2.12 ± 0.06	2.29 ±0.10	2.740	3.38
PDM1-2d	30.1	32.2	33.3	33.3	9.58 ± 0.10	9.70 ± 0.07	8.97 ± 0.14	0.052	1.75

Especially in modeling of $^{85}Sr^{2+}$ migration, both types of equilibrium equations can be used even if the model with incorporated Freundlich equation gives better results, i.e., the better goodness-of-fit. The model includes both partial processes, i.e. sorption and desorption. Concerning desorption, its efficiency very good in all cases. It can be explained as a result of presumed desorption mechanism that it is based on isotopic ion exchange between stable Sr^{2+} (SGW contained $10^{-7}M$ Sr^{2+}) and $^{85}Sr^{2+}$ sorbed on the surface of individual grains. This fact, of course, limits the general applicability of the obtained results. Therefore it implies the necessity of the studies under more realistic conditions, in particular, using a different type of SGW to the desorption process. From Tables 9 and 10 and especially from Fig. 24, it can be also seen that the values of R_S and R_D increase with decreasing grain size, i.e., with the increasing surface and with the increasing number of active sorption centers (sites) of both materials. However, R_S and R_D values are very close to each other for both rock types for corresponding grain size. Nevertheless, R_D for fracture infill material is about 3-times higher than that in pure granite, independently on the grain size, similarly as in the case of $^{137}Cs^+$ [24].

Table 10 The most important transport and sorption/desorption theoretical (calculated) parameters of $^{85}Sr^{2+}$ in crushed granite (PDM1-1a, 1b, 1c, 1d) and fracture infill (PDM1-2a, 2b, 2c, 2d). Non-linear isotherm approach (From Palágyi Š et al. (2013) J Radioanal Nucl Chem 297:33. With permission)

Columns	$R_{S,theor}$	$R_{D,theor}$	$(k_F)_S$	$(k_F)_D$	$(n_F)_S$	$(n_F)_D$	$Pe_{S,D}$	$D_{dS,D}$	WSOS/DF
	-	-	-	-	cm³/g	cm³/g	-	cm²/min	-
PDM1-1a	2.39 - 2.68	2.35 - 2.95	6.14 ± 0.75	0.21 ± 0.01	0.99 ± 0.01	0.87 ± 0.01	48.70 ± 0.39	8.77 E-3	6.97
PDM1-2a	56.2 - 56.8	51.8 - 57.0	28.01 ± 6.57	43.20 ± 6.15	1.01 ± 0.01	1.02 ± 0.01	25.70 ± 1.48	0.015	3.56
PDM1-1b	1.92 - 2.22	1.81 - 4.35	0.91 ± 0.70	0.015 ± 0.023	0.94 ± 0.02	0.80 ± 0.01	18.1 ± 0.30	0.024	4.06
PDM1-2b	54.2 - 62.2	63.9 - 76.2	45.00 ± 8.11	2.86 ± 0.35	1.03 ± 0.01	0.94 ± 0.01	42.10 ± 1.08	0.01	0.65
PDM1-1c	7.89 - 8.74	8.75 - 11.8	8.48E-5 ±5.24E-4	0.40 ± 0.20	0.65 ± 0.21	0.94 ± 0.02	2.58 ± 0.25	0.17	3.36
PDM1-2c	29.8 - 38.2	24.3 - 39.6	0.81 ± 0.08	7870 ± 1780	0.92 ± 0.01	1.22 ± 0.01	7.72 ± 0.04	0.06	0.40
PDM1-1d	4.25 - 4.91	5.37 - 8.47	3.72E-7 ±7.7E-7	131.00 ± 51.20	0.48 ± 0.07	1.14 ± 0.01	1.60 ± 0.08	0.29	3.13
PDM1-2d	29.12 - 34.4	30.2 - 38.6	1.75 ± 0.46	0.65 ± 0.14	0.95 ± 0.01	0.91 ± 0.01	9.41 ± 0.13	0.05	0.60

In the similar dependence of calculated values of K_{dS} and K_{dD} on the grain size, even closer equality between these values for the same average grain size can be shown. This may also confirm high reversibility of the sorption/desorption process in the investigated crushed rocks and sorption/desorption mechanism based on isotopic exchange of $^{85}Sr^{2+}$ with stable Sr^{2+}. In Fig. 24, the respective dependences were fitted with regression lines, which are presented in these figures. Exponential fit was applied, and for $R_{S,exp}$ and $R_{D,exp}$ values of coefficient of determination (R^2) between 0.82-0.89 and for $K_{dS,exp}$ and $K_{dD,exp}$, even higher coefficients of determination (0.90-0.97) were obtained, which also confirm the goodness of fit.

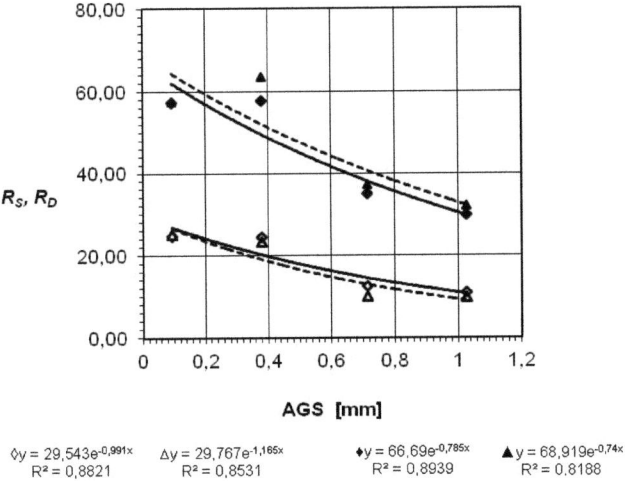

$\lozenge y = 29,543e^{-0,991x}$ $\triangle y = 29,767e^{-1,165x}$ $\blacklozenge y = 66,69e^{-0,785x}$ $\blacktriangle y = 68,919e^{-0,74x}$
$R^2 = 0,8821$ $R^2 = 0,8531$ $R^2 = 0,8939$ $R^2 = 0,8188$

Fig. 24 Effect of the average grain size (AGS) on the sorption retardation coefficients (see Table 3) R_S ($\lozenge$) and desorption retardation coefficients R_D ($\triangle$) of $^{85}Sr^{2+}$ in crushed granite and on the R_S ($\blacklozenge$) and R_D ($\blacktriangle$) of $^{85}Sr^{2+}$ in fracture infill, respectively. Symbols are experimental and lines theoretical values. (From Palágyi Š et al. (2013) J Radioanal Nucl Chem 297:33. With permission)

It was also interesting to observe the residual activity after $^{85}Sr^{2+}$ desorption. From the material balance of the activities flowing out of the columns and the residue in the columns after desorption, it was found that as low as about 2.4% of activity in crushed granite has remained permanently, i.e., irreversibly fixed in the column. In the case of fracture infill, the fixed activity in the column is even lower about 0.8%, only. These facts also confirm the practically complete reversible sorption/desorption of $^{85}Sr^{2+}$, which means that $^{85}Sr^{2+}$ reveals high mobility and can easily be transported in the studied crushed rocks by groundwater, if the conditions for isotope exchange mechanism exist (see the concentration Sr^{2+}). This result indicates the similarity with $^{137}Cs^+$, which desorption yield under the same conditions, from these rocks reached approx. 100%, especially from fracture infill [24].

2.2.2.4. Effect of grain size on the transport parameters of SeO_4^{2-} and SeO_3^{2-} in pure granite and fracture infill

Selenium (mainly ^{79}Se) has to be taken carefully into account in safety assessment of deep geological repository of radioactive waste due to its redox-sensitivity. Its mobility and occurrence depend on the actual oxidation state. In aqueous solutions it can occur in five oxidation states (-2, 0, +2, +4, +6) as anionic species. The species SeO_3^{2-} and SeO_4^{2-} are considered to be soluble mobile species. In the non-radioactive experiments sorption of Se(IV) and Se(VI), in $2 \cdot 10^{-5}$ mol/dm^3 Na_2SeO_4 and Na_2SeO_3 dissolved in synthetic granitic water (SGW), were investigated in columns of crushed granite and fracture infill (clay minerals) of various grain sizes. Desorption of these chemical forms was studied using SGW containing no selenium [34]. The goal of study was the quantification of the effect of grain size on the retardation, distribution and hydrodynamic dispersion coefficients of SeO_4^{2-} and SeO_3^{2-}.

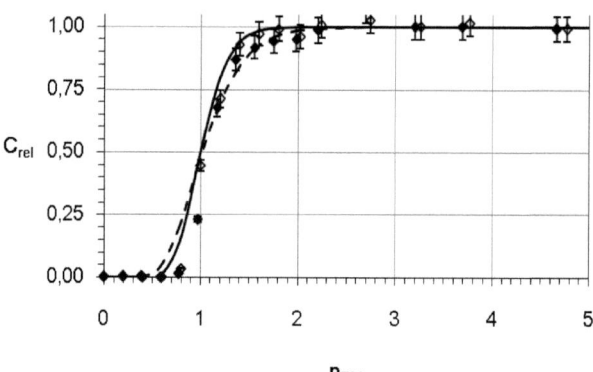

Fig. 25 Sorption BTCs of SeO_4^{2-} in granite (◊) and in fracture infill (♦) (symbols: experimental values, curves: calculated values; solid lines – to (◊), dotted lines – to (♦) (From Videnská K et al. (2013) J Radioanal Nucl Chem 298:547. With permission)

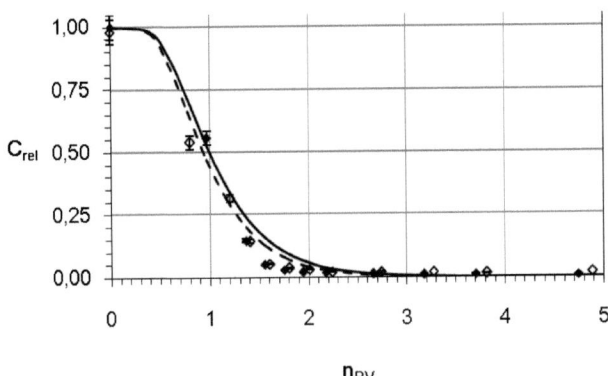

Fig. 26 Desorption BTCs of SeO_4^{2-} in granite (◊) and fracture infill (♦). (Legend: Fig. 1) (From Videnská K et al. (2013) J Radioanal Nucl Chem 298:547. With permission)

Crystalline rock materials and synthetic groundwater were the same as in subchapter 2.2.2.2 and 2.2.2.3. Experimental conditions were identical to the conditions of experiments with Cs^+ or Sr^{2+} presented in papers [24,25] and in mentioned subchapters. Solutions of Na_2SeO_4 and Na_2SeO_3 in SGW were added individually into a suitable reservoir. The determination of selenium in initial and in collected fractions at the 1.3 cm inner diameter and 5.4 cm column outflow was performed using the ICP-MS technique. The most abundant isotope [80]Se was used for the measurement and the spectral interferences due to [40]Ar_2^+ were removed by methane as reaction gas and Ge served as the internal standard. SGW flow-rate was ca. 0.06 cm^3/min and the pore volume was about 3 cm^3. After sorption, desorption was carried out with SGW without selenium presence. The experiments with selenate and selenite were preceded by experiments with ^{3}H, which served for the evaluation of the flow conditions.

Sorption and desorption BTC for SeO_4^2 for grain sizes between 0.063-0.125 mm are presented in Figs. 25 and 26. From the evaluation of sorption and desorption breakthrough curves of all grain size fractions it follows that retardation coefficients are close to 1 (i.e., both n_{PVSi} and n_{PVDi} are approximately equal to 1 at $C_{rel} = 0.5$) for each investigated grain sizes, as seen in Table 11. It means that both experimental and calculated (theoretical) K_d values converge to 0. Let us admit that from theoretical viewpoint, the retardation coefficient cannot acquire lower values than 1, as it follows from the relation of $R=1+\rho \cdot K_d$ $/\theta$. Therefore it is evident, that the deviances in R_{expS} and R_{expD} values from 1 are a consequence of experimental failure. From Table 11 it is clear that SeO_4^{2-} behaves as conservative tracer, which does not sorb on the surface of granite or fracture infill.

Breakthrough curves of SeO_3^{2-} sorption and desorption in both granitic materials are demonstrated in Figs. 27 and 28. Similarly, like in the SeO_4^{2-} sorption, theoretical breakthrough curves fit the experimental values satisfactorily well also in SeO_3^{2-} sorption. Sorption and transport parameters of this study are presented in Table 12. On the contrary to SeO_4^{2-}, sorption of SeO_3^{2-} in these materials is evident. Especially in the fracture filling material, the selenite is rather strongly retarded due to pronounced sorption. Further, it can be seen that the grain size has a significant effect on both the R_{exp} and R_{theor} values, as well as on the K_{dexp} and K_{dtheor} values. It is quite clear, that decreasing in the grain size causes an increase in the sorption of SeO_3^{2-}. It means that the smaller is the grain size, the higher is the retardation of selenite, more for fracture infill and less for pure granite. The fact that the respective K_{dexpD} values are lower than K_{dexpS} values signalizes that a certain concentrating of SeO_3^{2-} takes place especially on the fracture infill, which is visible from the comparison of breakthrough curves of sorption and desorption (Figs. 27 and 28).

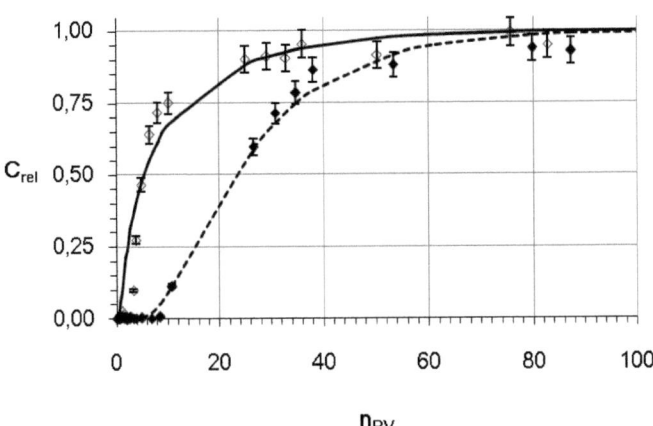

Fig. 27 Sorption BTC of SeO_3^{2-} in granite ($\Diamond$) and in fracture infill ($\blacklozenge$). (Legend: Fig. 1) (From Videnská K et al. (2013) J Radioanal Nucl Chem 298:547. With permission)

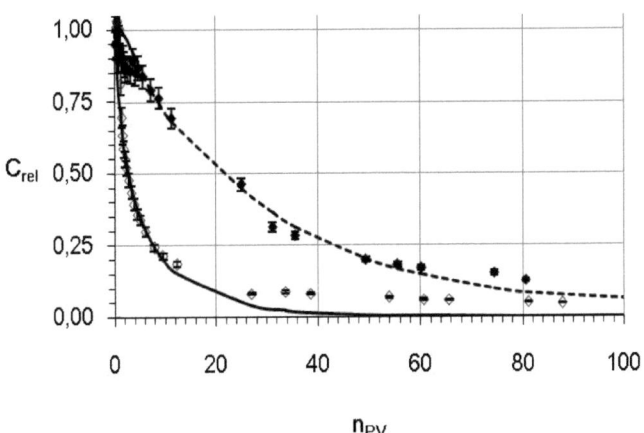

Fig. 28 Desorption BTC of SeO_3^{2-} in granite ($\Diamond$) and in fracture infill ($\blacklozenge$). (Legend: Fig. 1) (From Videnská K et al. (2013) J Radioanal Nucl Chem 298:547. With permission)

It is necessary to be noted that in the case of all experiments with SeO_4^{2-} and SeO_3^{2-}, values of Freundlich parameter n converged to unity and therefore only values of K_d can be found in tables. Both parameters have similar values in a case of desorption which reflects the reversible character of sorption process. It was found that besides retardation and distribution coefficients and sorption capacity for SeO_3^{2-} also increases with decreasing grain size.

Table 11 The most important transport and sorption/desorption parameters of SeO_4^{2-} in crushed granite (PDM1-1a, -1b, -1c, -1d) and fracture infill (PDM1-2a, -2b, -2c, -2d) (From Videnská K et al. (2013) J Radioanal Nucl Chem 298:547. With permission)

Column	R_{expS}	R_{expD}	R_{teorS}	R_{teorD}	K_{dteorS}	K_{dteorD}	$Pe_{S,D}$	$D_{dS,D}$	WSOS/ DF
code	-	-	-	-	cm^3/g	cm^3/g	-	cm^2/min	-
PDM1-1a	1.04	0.98	1.00	1.00	0.0004	0.00004	5.0	0.109	>50 (?)
PDM1-2a	1.09	0.83	1.01	1.00	0.0044	0.00013	4.98	0.101	>50 (?)
PDM1-1b	0.98	0.99	1.03	1.00	0.0082	0.00001	18.2	0.030	4.27
PDM1-2b	0.89	0.95	1.12	1.26	0.0330	0.09300	19.3	0.029	1.17
PDM1-1c	0.89	0.92	1.00	1.00	0.0014	0,00001	47.3	0.011	>50 (?)
PDM1-2c	0.85	1.02	1.00	1.01	0.0002	0.00040	41.1	0.014	7.71
PDM1-1d	0.88	0.98	1.42 (?)	1.02	0.13(?)	0.00500	23.0	0.019	19.7
PDM1-2d	0.85	0.90	1.04	1.00	0.0120	0.00001	41.4	0.011	17.3

Table 12 The most important transport and sorption/desorption parameters of SeO_3^{2-} in crushed granite (PDM1-1a, -1b, -1c, -1d) and fracture infill (PDM1-2a, -2b, -2c, -2d) (From Videnská K et al. (2013) J Radioanal Nucl Chem 298:547. With permission)

Column	R_{expS}	R_{expD}	R_{teorS}	R_{teorD}	K_{dteorS}	K_{dteorD}	$Pe_{S,D}$	$D_{dS,D}$	WSOS/ DF
code	-	-	-	-	cm^3/g	cm^3/g	-	cm^2/min	-
PDM1-1a	4.67	2.49	8.70	4.64	2.55	1.21	0.5(?)	1.18	30.5
PDM1-2a	20.85	16.85	37.20	30.06	12.70	10.10	5.0	0.11	10.0
PDM1-1b	3.22	1.47	3.33	2.20	0.67	0.34	7.0	0.07	13.2
PDM1-2b	6.08	1.39	4.62	1.73	1.01	0.21	2.4	0.23	8.63
PDM1-1c	1.41	1.42	1.58	1.67	0.17	0.20	10.4	0.05	14.1
PDM1-2c	2.35	1.07	2.68	1.49	0.50	0.15	7.1	0.07	5.47
PDM1-1d	2.11	1.51	1.58	1.14	0.17	0.05	1.1	0.37	2.43
PDM1-2d	2.10	1.04	1.81	1.01	0.24	0.00	0.7	0.63	1.37

2.2.2.5. Interaction of $^3H^+$ and $^{36}Cl^-$ with pure granite and fracture infill materials

Another example of the effectiveness of the used fitting procedure using Eq. (16) have been reported in the investigation of $^3H^+$ (as HTO) and $^{36}Cl^-$ (as $Na^{36}Cl$) interaction with crushed pure granite and corresponding fracture infill material in column experiments during their transport by groundwater [35].

Results of column experiments with these radionuclides can provide information about the properties of the solid phase in the column. Namely, the values of Peclet number, longitudinal hydrodynamic dispersion coefficient and accessible porosity can be obtained from the peak and/or from the whole BTC. Radionuclides ^{3}H (in the form of tritiated water, HTO) and ^{36}Cl⁻ (in an anionic form Na^{36}Cl) are regarded as non-interacting ($K_d \rightarrow 0$, $R \rightarrow 1$) conservative tracers. Moreover, anionic exclusion is usually under consideration for ^{36}Cl⁻.

On the other hand, dynamic experiments with these radionuclides can be used for the hydrodynamic characterization of the given column [36,37]. Experiments with natural fracture column showed higher retardation and dispersion of tritiated water in comparison with chlorine, probably due to ion-exclusion of the anions in fracture infill [38]. Effect of matrix diffusion on transport of chlorine and tritium was observed in case of lower flow rates. Advection was dominant process at the fastest flow rates of liquid phase. Our experiments were focused on the study of the effect of grain size, firstly on the values of Peclet number, longitudinal dispersion coefficient and bed porosity, secondly, on the influence of grain size on the values of retardation and distribution coefficients.

Analogous to the previous studies reported in subchapters 2.2.2.2 – 2.2.2.4, two types of crystalline rocks were used: pure granite, coded as PDM1-1, and fracture infill material, coded as PDM1-2. The column sizes, composition of SGW and pore-water volume were same as in previous subchapters. The activity on the input and flow-rate of SGW (about of 0.05 cm^3/min) through the columns were constant during the sorption process.

The experimental BTCs of ^{3}H$^+$ (as HTO) and ^{36}Cl⁻ (as Na^{36}Cl) obtained both with crushed granite and their infill materials, as well as the results of the fitting of these data with theoretical BTC calculated by linear isotherm approach model, are demonstrated for grain size 0.063 – 0.125 mm in Figs. 29 and 30 for sorption and desorption of ^{3}H$^+$, respectively, and in Figs. 31 and 32 for sorption and desorption of ^{36}Cl⁻, respectively. The interaction of components studied with solid phase is weak, as it is evident from comparison the values of R_{Scal} and $K_{d,Scal}$, the linear isotherm approach proved to be quite suitable. The breakthrough curves for other grain sizes are similar or practically the same. This similarity is evident from Table 13, where the important transport parameters are summarized, including the values of goodness-of-fit criterion declaring the very good agreement between experimental and calculated data.

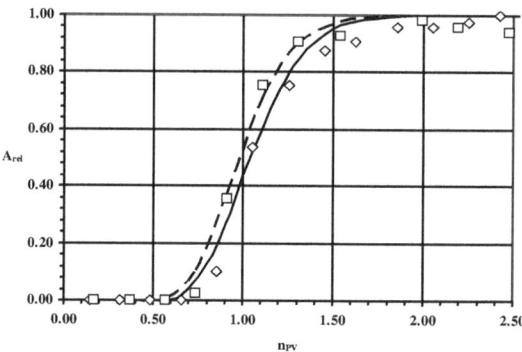

Fig. 29 Sorption BTC for $^3H^+$ (as HTO) in crushed pure granite ($\Diamond$) and fracture infill ($\square$). Symbols: experimental data, lines: calculated values (solid line to ($\Diamond$), dash line to ($\square$) (From Štamberg K et al. (2014) J Radioanal Nucl Chem 299:1625. With permission)

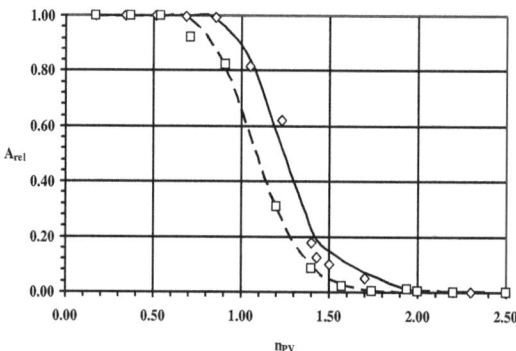

Fig. 30 Desorption BTCs for $^3H^+$ (as HTO) in crushed pure granite ($\Diamond$) and fracture infill ($\square$). Legends see Fig. 1 (From Štamberg K et al. (2014) J Radioanal Nucl Chem 299:1625. With permission)

Table 13 The most important transport parameters of $^3H^+$ and $^{36}Cl^-$ during their sorption in crushed pure granite and its fracture infill (From Štamberg K et al. (2014) J Radioanal Nucl Chem 299:1625. With permission)

Nuclide	Rock	R_{Sexp} -	R_{Scal} -	$K_{d,Scal}$ g cm^{-3}	Pe -	D_d cm^2 min^{-1}	$WSOS$ /DF
$^3H^+$	granite	1.13	1.05	0.01	74.3	0.006	0.86
	infill	0.91	1.02	0.01	46.5	0.009	2.08
$^{36}Cl^-$	granite	1.08	1.03	0.00	27.4	0.018	1.25
	infill	0.86	1.00	0.00	31.4	0.016	0.10

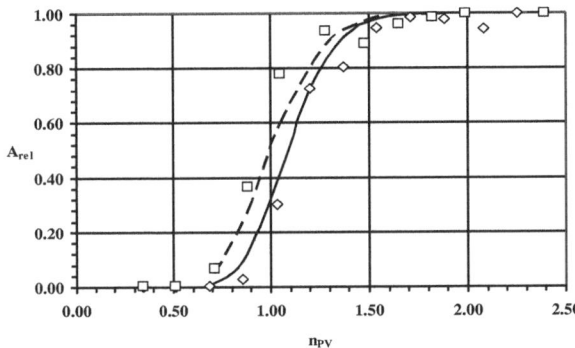

Fig. 31 Sorption BTCs for $^{36}Cl^-$ (as NaCl) in crushed pure granite (◊) and infill (□) material. Symbols: experimental data, lines: calculated values (solid line – to (◊), dash line – to (□) (From Štamberg K et al. (2014) J Radioanal Nucl Chem 299:1625. With permission)

With a few exceptions, the values of theoretical R_{exp} were practically equal to 1 and the $K_{d,exp}$ values converged to zero in case of all fractions of crushed granite as well as of fracture infill material. This means that no noticeable interaction of these radionuclides was found with studied rock materials, only a very weak ion exclusion of $^{36}Cl^-$ was determined in the case of infill material. Also, the influence of grain size on R was not observed. Practically, tritium and chloride behaved here as non-interacting, conservative tracers. Different pattern was observed in case of Peclet number and dispersion coefficient. Generally, Pe numbers for T and Cl decreased (and logically D_d increased) with increasing grain size of granite and fracture infill. But, the dependences did not agree with the theoretical suppositions in all cases. The differences could be caused, e.g., by the size and shape distribution of particles in bed, by the non-homogeneity of samples and the non-uniform bed porosity resulting in different tortuosity of the path of given tracer in the bed. Generally, the flow pattern can be different even if the column media originates from the one borehole.

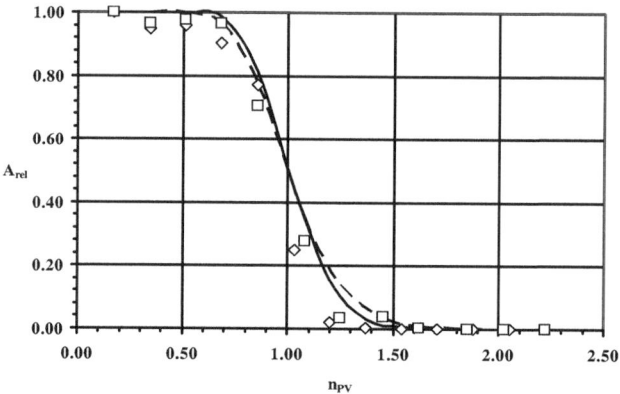

Fig. 32 Desorption curves for $^{36}Cl^-$ (as NaCl) in crushed pure granite ($\Diamond$) and in infill ($\square$). Symbols: experiment, lines: calculated values (solid line to ($\Diamond$), dash line to ($\square$) (From Štamberg K et al. (2014) J Radioanal Nucl Chem 299:1625. With permission)

2.2.2.6. Transport parameters of I^- and IO_3^- determined in granitic rocks

^{129}I is radiologically a very important nuclide also for deep geological repository of radioactive wastes mainly due to its long half-life of radioactive decay ($15.7 \cdot 10^6$ years). Also, appreciable amounts of ^{129}I have been found in soil all over the world. Radioiodine exhibits a vast number of physical and chemical forms in valence states from 1- to 7+ (I^-, I_2, I_3^-, IO^-, IO_2^-, IO_3^-, IO_4^-, etc.), including organic compounds, encountered in the technological systems of a nuclear power plants, reprocessing equipments, or even radioactive disposal facilities and their surrounding environment. Recently, some differences have been observed in the sorption behavior of species with different valence states of iodine, namely its anions, in various rocks, sediments and soils under static as well as column conditions [30,40]. Therefore it seemed interesting to investigate the transport of two principal inorganic iodine anions, simple iodide anion and iodate oxoanion, in crushed crystalline granitic rocks of domestic origin. Two types of crystalline granitic rocks were used in our study [39]: gabbro and tonalite, which originated from the locality mentioned in subchapter 2.1.2.1. Each rock sample was crushed and dry sieved to several grain fractions and a homogenized fraction of 0.25 – 0.80 mm was used in the experimental study. Instead of ^{129}I radioiodine stable iodine compounds were used, as carrier free radioiodate (e.g., $^{131}IO_3^-$ or $^{125}IO_3^-$ substitute) is not readily available. The column sizes, composition of SGW and pore-water volume were same as in previous subchapter. Throughout the experiments a flow-rate of 0.06±0.01 cm^3/min was applied. The pore water volume was about 3 cm^3.

Both chemical forms were determined by UV-VIS spectrophotometric technique (SPECORD® PC 205 UV-VIS Spectrophotometer (Analytica, Jena, Germany)) after their transformation to I_2 using appropriate amplification reactions [41,42]. In the determination of iodide the following simple oxidation reaction were used:

$$2\ I^- + NO_2^- + 4\ H^+ = I_2 + 2\ NO + 2\ H_2O, \tag{18}$$

and for iodate determination the reaction:

$$IO_3^- + 5\ I^- + 6\ H^+ = 3\ I_2 + 3\ H_2O. \tag{19}$$

The second reaction is a well-known Dushman reaction or Leipert reaction of amplification nature and therefore this reaction enabled the iodate determination with 3-times higher sensitivity than iodide. In both determinations the liberated I_2 was extracted into $CHCl_3$ and resulted in the violet color. For a given distribution coefficient and a phase ratio only 0.8% of iodine remains in the aqueous phase after extraction. According to the manufacturer data the absorption maximum for I_2 at concentrations of 10^{-3}-10^{-4} mol/cm^3 lies between 508-540 nm.

On the basis of corresponding BTCs (not presented here) it was found that no of these species of iodine (1- and 5+) were sorbed on the investigated granites, what is seen from the almost identical values of retardation coefficients presented in Table 14. The iodine species behaved as conservative tracers and therefore, the linear isotherm approach was believed to be suitable to the modeling of BTCs. On the other hand, the values of *WSOS/DF* point at the relatively high experimental error probably due to the change of liquid phase flow-rate, flow pattern and channeling in column.

Theoretical values of the retardation coefficients, namely of $R_{S,T}$ and $R_{D,T}$) of about 1 have been obtained for both I^- and IO_3^-, which show practically non-sorptive behavior of these chemical forms of iodine in the tested granitic rocks or it means that the theoretical values of distribution coefficients converge to zero. But, the experimental values of retardation coefficients (Table 14), especially in a case of IO_3^-, are smaller than 1, which leads to a conclusion that the ion exclusion can be supposed. Such supposition corresponds well with the results concerned of the behavior of anionic forms in the course of dynamic sorption on granitic materials published in [38]. These results have shown that the seepage velocity of iodine species (u_I) can be equal, or greater than seepage velocity of SGW (u_w):

$$u_I = R_{S,D} \cdot u_w \qquad (20)$$

Table 14 Transport parameters of two principal iodine species in rock samples (From Palágyi Š and Štamberg K (2014) J Radioanal Nucl Chem 302:647. With permission)

Granitic rocks	Iodine species	$R_{S,E}$	$R_{S,T}$	$R_{D,E}$	$R_{D,T}$	Pe	D_d	WSOS/
		-	-	-	-	-	cm^2/min	DF
Gabbro	I^-	1.14	1.02±0.01	0.89	1.00±0.01	21.9±3.3	0.0079	17.8
	IO_3^-	0.97	1.00±0.01	0.96	1.00±0.01	20.0±0.9	0.0078	17.1
Tonalite	I^-	1.05	1.13±0.04	0.89	1.00±0.01	24.3±1.6	0.0067	13.0
	IO_3^-	0.88	1.00±0.01	0.88	1.00±0.01	20.0±1.8	0.0084	14.2

It means that the iodine species are not retarded at all in their transport through investigated crushed granitic rocks with granitic water. Under given conditions the values of hydro-dynamic dispersion coefficient were about 0.0077 ± 0.0006 cm^2/min for both anions and rocks. These experiments indicate that the host rock based on granite of the radioactive waste repositories may not act as a natural barrier against transport of anionic radioiodine nuclides, if they reach the groundwater. For the better insight into the sorption problem, static batch experiments were carried out at a liquid/solid ratio of 3 confirm the above findings, in which only as low as 9 ± 4 % of iodide and $8 \pm 3\%$ of iodate were sorbed during 3 days in both rocks. A relatively higher sorption under static condition may be due to longer contact time in comparison with dynamic conditions. It can also be ascribed to a slight oxidation of iodide and/or reduction of iodate to I_2, which sorption in rocks is more enhanced than the anionic species. This may caused by the presence of some minor amount organic compounds, which were found in granite. The sorption is, highly probable, rate limited.

For their non-sorptive behavior, neither iodide nor iodate is retained, or at least retarded, in the studied crushed crystalline granites during their transport from granitic groundwater. The results suggest that practically no difference should be expected in the transport and/or retardation of these two possible species of water-borne radioiodine in gabbro or tonalite, under investigated or similar conditions. These experiments indicate that the host rock based on granite of the radioactive waste repositories may not act as a natural barrier against transport of anionic radioiodine nuclides, if they reach the groundwater. The definite appraisal of the behavior of inorganic anionic species of iodine in such types of granitic rock would require further investigation of the possible sorption in rocks crushed of smaller grain size, when next, additional more active centers may be uncovered. Nevertheless, from our results, as well as from the supporting literature data, it can be concluded that the oxidation of iodide to iodate gives only limited solution of ^{125}I$^-$ iodine retaining from the outer environment.

2.2.2.7. Determination of sorption capacity of fucoidic sands for Cs$^+$ and Sr^{2+}

The determination of the capacity of fucoidic sands for univalent and divalent cations is a great importance in remediation processes. The capacity can be defined in numerous ways, of which mainly ion exchange capacity, or sorption capacity seems to be useful [44]. One of the positive features of dynamic column technique is its ability to determinate of so-called breakthrough capacity. In this work, radioactive indicators ^{137}Cs and ^{85}Sr added to the inactive Cs and Sr salts dissolved in Cenoman background water has been used for the determination of the Cs and Sr sorption capacity of fucoidic sands. As column beds, intact undisturbed samples of drilled fucoidic sandstones were used in study described in [43], same as described in subchapter 2.2.2.1 along with other technical details. The working procedure also was the same.

The fitting BTCs of experimental data are presented in Figs. 13 and 14 in the subchapter 2.2.2.1. Only ascending (sorption) parts of the curves were used for the determination of the sorption capacity. The numerical results are given in Table 15. Sorption capacities were calculated from theoretical BTCs agreeing well with respective experimental BTCs. Sorption capacity (SC) was calculated for 5, 10, 50, 90, 95 and 100% breakthrough from the standardized activity values of SGW and the volume of SGW passed through the column according following simple relation:

$$SC = V_{n\%} \cdot C \cdot 100/G \text{ [mol/100 g]}, \tag{21}$$

where: $V_{n\%}$ - volume of SGW (cm^3) at n % breakthrough ($n = 5-100\%$), C – concentration of the element (mol/ dm^3) and G – mass of the sand cylinder (g).

From the results it follows that values of the sorption capacity of fucoidic sands for Cs$^+$ a Sr^{2+} under dynamic conditions are relatively low and reach fractions of µmol/100g, whereas values of the sorption capacity in the same samples for Cs^+ are higher than for Sr^{2+}. By comparison of values for Cs^+ obtained with different methods it is evident that the values of sorption capacity under dynamic conditions are lower of about one order of magnitude than under static (batch) conditions.

Table 15 Experimental (exp) and calculated theoretical (thr) sorption capacity of fucoidic sands for Cs and Sr (From Palágyi Š et al. (2010) J Radioanal Nucl Chem 286:317. With permission)

Break-through	48118Cs				48562Cs			
	SGW cm³		Sorption capacity µmol/100g		SGW cm³		Sorption capacity µmol/100g	
%	exp	thr	exp	thr	exp	thr	exp	thr
5	1412	1328	0.38	0.36	285	219	0.08	0.06
10	1517	1427	0.41	0.39	339	274	0.09	0.08
50	1874	1840	0.51	0.50	614	599	0.17	0.17
90	4957	4063	1.34	1.10	1201	1064	0.33	0.29
95	5314	4550	1.44	1.23	1534	1242	0.42	0.34
100	5741	5703	1.56	1.55	1828	1755	0.50	0.48
	48511Sr				58562Sr			
	SGW cm³		Sorption capacity µmol/100g		SGW cm³		Sorption capacity µmol/100g	
5	423	395	0.12	0.11	127	120	0.4	0.3
10	465	453	0.13	0.13	148	144	0.4	0.4
50	735	735	0.20	0.20	253	253	0.7	0.7
90	1273	1228	0.35	0.34	512	449	0.14	0.13
95	1484	1485	0.41	0.41	556	535	0.15	0.15
100	1749	1760	0.49	0.49	1166	989	0.32	0.28

3. Conclusions

The column technique combined with simplified dynamic modeling proved to be successful in the determination of basic transport parameters in crushed rock-groundwater and/or soil-groundwater systems. This technique can be used in the both basic laboratory methods, i.e., of pulse tracer inlet and step (continuous) trace inlet of liquid phase into the given column. Due to the fact that the models used for the evaluation of column experiments are based on an analytical solution of interaction-advection-dispersion equation, the experimentally data, can be easily evaluated not only in the case of linear sorption/desorption isotherm equilibrium approach, but also of non-linear isotherm approach. In the overview of our study of different sorbing species through the bed of rock material it was generally shown, that the values of *WSOS/DF* in the non-linear isotherm sorption approach are lower in comparison with the linear isotherm sorption approach. Only in cases, in which the interaction of contaminants with the solid phase is weak, or on the other hand very strong ($K_d > 10^3 - 10^4$), the linear isotherm approach can be characterized as satisfactory. As can be seen in the presented figures, the result of fitting in step (continuous) inlet is sufficiently close to experimental points and does not need any further correction as the pulse inlet technique. Therefore, the step tracer inlet technique seems to be better than the pulse inlet, especially as a method aimed at the determination of transport parameters on a laboratory scale. The technique can be also applicable to systems with granular matter of impaired homogeneity, yet the breakthrough curves may be partially distorted. The step inlet technique itself gives somewhat better results; however the pulse inlet technique of radioactive contaminants may save radioactive materials and improve radiation safety in the experimental studies. In addition to the advantages above mentioned, the characteristic features of simplified dynamic modeling are the easy construction of corresponding codes, fast solubility of model equations, including non-linear regression procedure, and as a result very short time needed for the evaluation of experimental data. Furthermore it enables, among other things, to obtain the information about the type of equilibrium isotherm and the concentration dependence of retardation coefficient.

4. References

1. IAEA (1985) Deep Underground Disposal of Radioactive Wastes: Near-Field Effects. Technical Report Series No. 251, Vienna

2. Alexander WR, Smith PA, McKinley IG (2003) Modelling radionuclide transport in the geological environment, in Modelling Radioactivity in the Environment, Scott EM (Ed) Elsevier, Amsterdam, p. 109-145

3. Barnett MO, Jardine PM, Brooks SC, Selim HM (2000) Adsorption and transport of Uranium (VI) in subsurface media. Soil Sci Soc Am J 64: 908-917

4. Szenknect S, Ardois C, Gaudet JP, Barthes V (2005) Reactive transport of ^{85}Sr in a Chernobyl sand column: static and dynamic experiments and modeling, J Contam Hydrol 76:139-165

5. Melkior T, Yahiaoui S, Motellier S, Thoby D, Tevissen E (2005) Cesium sorption and diffusion in Bure mud rock samples. Applied Clay Science 29:172-1860

6. Palágyi Š, Vodičková H (2009) Sorption and desorption of ^{125}I, ^{137}Cs$^+$, ^{85}Sr^{2+} and ^{152,154}Eu^{3+} on disturbed soils under dynamic flow and static batch conditions, J Radioanal Nucl Chem, 280:3-14

7. Palágyi Š, Laciok A (2006) Sorption, desorption and extraction of uranium from some sands under dynamic conditions, Czechoslov J Phys, 56:D483-D492

8. Palágyi Š, Vodičková H, Landa J, Palágyiová J, Laciok A (2009) Migration and sorption of ^{137}Cs and ^{152,154}Eu in crushed crystalline rocks under dynamic conditions. J Radioanal Nucl Chem 279:431-441

9. Palágyi Š, Štamberg K, Vodičková H, Hercík M (2013) Sorption of ^{125}I, ^{137}Cs$^+$, ^{85}Sr^{2+} and ^{152,154}Eu^{3+} during their transport in undisturbed vertical and horizontal soil cores under dynamic flow conditions. J Radioanal Nucl Chem 295:1447-1458

10. IAEA (2003) Scientific and Technical Basis for Geological Disposal of Radioactive Wastes. Technical Report Series No. 413, Vienna.

11. Mell P, Megyeri J, Riess L, Máthé Z, Hámos G, Lázár K (2006) Diffusion of Sr, Cs, Co and I in argillaceous rock as studied by radiotracers. J Radioanal Nucl Chem 268:411-417

12. Xu Zhe, Cai Jian-Guo, Pan Bing-cai (2013) Review: Mathematically modeling fixed-bed adsorption in aqueous systems. J Zhejiang Univ-Sci A (Appl Phys & Eng) 14(3):155-176

13. Riazi M, Kesthkar AR, Moosavian MA (2014) Batch and continuous fixed-bed column biosorption of thorium (IV) from aqueous solutions: equilibrium and dynamic modeling. J Radioanal Nucl Chem 301:493-503

14. Kumar A, Rout S, Chopra MK, Mishra DG, Singhal RK, Ravi PM, Tripathi RM (2014) Modeling of ^{137}Cs migration in cores of marine sediments of Mumbai Harbor Bay. J Radioanal Nucl Chem 301:615-626

15. Likar A, Omahen G, Lipoglavsek M, Vidmar T (2001) A theoretical description of diffusion and migration

of ^{137}Cd in soil. J Environ Radioactivity 57:191-201

16. Ozdural AR, Alkan A, Kerkhof PJAM (2004) Modeling chromatographic columns. Non-equilibrium packed-bed adsorption with non-linear adsorption isotherms. J Chromatography A 1041:77-85

17. Ebert K, Ederer H (1985) Computeranwendungen in der Chemie. VCH Verlagsgesellschaft mbH, Weinheim

18. Dvořák L, Ledvinka M, Sobotka, M (1991) Famulus 3.1, Computer Equipment, Prague

19. Herbelin AL, Westall JC (1996) FITEQL – A Computer Program for Determination of Chemical Equilibrium Constants from Experimental Data, Version 3.2., Report 94-01, Department of Chemistry, Oregon State University, Corvallis, Oregon

20. Palágyi Š, Štamberg K (2010) Modeling of transport of radionuclides in beds of crushed crystalline rocks under equilibrium non-linear sorption isotherm conditions. Radiochim. Acta 98:359-365

21. Palágyi Š, Štamberg K, Vodičková H (2010) Transport and sorption of ^{85}Sr and ^{125}I in crushed crystalline rocks under dynamic flow conditions. J Radioanal Nucl Chem 283:629-636

22. Palágyi Š, Štamberg K (2010) Transport of ^{125}I$^-$, ^{137}Cs$^+$ and ^{85}Sr^{2+} in granitic rock and soil columns. J Radioanal Nucl Chem 286:309-316

23. Palágyi Š, Štamberg K (2011) Determination of ^{137}Cs and ^{85}Sr transport parameters in fucoids sand columns and groundwater system. Cent Eur J Chem 9:798-807

24. Štamberg K, Palágyi Š (2011) Effect of grain size on the sorption and desorption of ^{137}Cs in crushed granite columns and groundwater system under dynamic conditions. J Radioanal Nucl Chem 293:127-134

25. Palágyi Š, Štamberg K, Havlová V, Vodičková H (2013) Effect of grain size on the ^{85}Sr^{2+} sorption and desorption in columns of crushed granite and infill materials from granitic water under dynamic conditions. J Radioanal Nucl Chem 297:33-39

26. Palágyi Š, Štamberg K, Vopálka D (2015) A simplified approach to evaluation of column experiments as a tool for determination of radionuclide transport parameters in rock-groundwater or soil-groundwater systems. J Radioanal Nucl Chem 304:945-954

27. Rachinskiy BV (1964) Vvedenije v obščuju teoriju dinamiky sorbcii i chromatografii (Introduction into General Theory of the Dynamics of Sorption and Chromatography), Nauka, Moskva.

28. Flury M, Czigány Sz, Chen G, Harsh JB (2004) Cesium migration in saturated silica sand and Hanford sediments as impacted by ionic strength. J Contam Hydrol 71:111-126

29. Braida WJ, Pignatello JJ, Lu Y, Ravikovitch PI, Neimark AV, Xing B (2003) Sorption hysteresis of benzene in charcoal particles. Environ sci technol 37:409-417

30. Hu Q, Zhao P, Moran JE, Seaman JC (2005) Sorption and transport of iodine species in sediments from the Savannah River and Hanford Sites Iodine is an important element in studies of environmental protection. J Contam Hydrol 78:185-205

31. Zheng C, Bennett GD (1995) Applied Contaminant Transport Modeling: Theory and Practice. Van Nostrand Reinhold, New York, p. 440

32. Van Genuchten MT, Wierenga PJ (1976) Mass Transfer Studies in Sorbing Porous Media I, Analytical Solutions. Soil Sci Soc Am 40:473- 480

33. Bear J, Verruijt A (1987) Modelling Groundwater Flow and Pollution: Theory and Applications of

Transport in Porous Media, D. Riedel Publishing Co., Dodrecht

34. Videnská K, Palágyi Š, Štamberg K, Vodičková H, Havlová V (2013) Effect of grain size on the sorption and desorption of SeO_4^{2-} and SeO_3^{2-} in columns of crushed granite and fracture infill from granitic water under dynamic conditions. J Radioanal Nucl Chem 298:547-554

35. Štamberg K, Palágyi Š, Videnská K, Havlová V (2014) Interaction of $^3H^+$ (as HTO) and $^{36}Cl^-$ (as $Na^{36}Cl$) with crushed granite and corresponding fracture infill material investigated in column experiments. J Radioanal Nucl Chem 299:1625-1633

36. Bazer-Bachi F, Descostes M, Tevissen E, Meier P, Grenut B, Simonno MO, Sardin M (2007) Characterization of sulfate sorption on Callovo - Oxfordian argillites by batch, column and through-diffusion experiments. Phys Chem Earth 32:552–558

37. Missana T, Alonso U, Garcia-Gutierrez M, Mingarro M (2008) Role of bentonite colloids on europium and plutonium migration in a granite fracture. Appl Geochem 23:1484–1497

38. Höltta P, Siitari-Kauppi M, Hamanen M, Huitti T, Hautojarvi A, Lindberg A (1997) Radionuclide transport and retardation in rock fracture and crushed rock column experiments. J Contam Hydrol 26:135–145

39. Palágyi Š, Štamberg K (2014) Transport parameters of I^- and IO_3^- determined in crushed rock column and groundwater system under dynamic flow conditions. J Radioanal Nucl Chem 302:647-653

40. Muramatsu Y, Uchida S, Sriyotha P, Sriyotha K (1990) Some considerations on the iodine sorption and desorption phenomena of iodine and iodate on soil. Water Air Soil Pollut 49:125–138

41. Belcher R (1968) Amplification reactions. Talanta 15:137–166

42. Furuichi R, Matsuzaki I, Simic R, Liebhafsky HA (1972) Rate of Dushman reaction at low iodide concentrations. Inorg Chem 11:952–955

43. Palágyi Š, Franta P, Vodičková H (2010) Determination of sorption capacity of fucoidic sands for Cs^+ and Sr^{2+} under dynamic column conditions. J Radioanal Nucl Chem 286:317–322

44. Appello CGJ, Postma D (1993) Geochemistry, Groundwater and Pollution, Balkema, Rotterdam, 535 pp

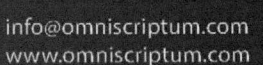

Printed by Books on Demand GmbH, Norderstedt / Germany